T0318380

OQAM/FBMC FOR FUTURE WIRELESS COMMUNICATIONS

OQAM/FBMC FOR FUTURE WIRELESS COMMUNICATIONS
Principles, Technologies and Applications

TAO JIANG

DA CHEN

CHUNXING NI

DAIMING QU

Academic Press is an imprint of Elsevier
125 London Wall, London EC2Y 5AS, United Kingdom
525 B Street, Suite 1800, San Diego, CA 92101-4495, United States
50 Hampshire Street, 5th Floor, Cambridge, MA 02139, United States
The Boulevard, Langford Lane, Kidlington, Oxford OX5 1GB, United Kingdom

© 2018 Elsevier Ltd. All rights reserved.

No part of this publication may be reproduced or transmitted in any form or by any means, electronic
or mechanical, including photocopying, recording, or any information storage and retrieval system,
without permission in writing from the publisher. Details on how to seek permission, further
information about the Publisher's permissions policies and our arrangements with organizations such as
the Copyright Clearance Center and the Copyright Licensing Agency, can be found at our website:
www.elsevier.com/permissions.

This book and the individual contributions contained in it are protected under copyright by the
Publisher (other than as may be noted herein).

Notices
Knowledge and best practice in this field are constantly changing. As new research and experience
broaden our understanding, changes in research methods, professional practices, or medical treatment
may become necessary.

Practitioners and researchers must always rely on their own experience and knowledge in evaluating
and using any information, methods, compounds, or experiments described herein. In using such
information or methods they should be mindful of their own safety and the safety of others, including
parties for whom they have a professional responsibility.

To the fullest extent of the law, neither the Publisher nor the authors, contributors, or editors, assume
any liability for any injury and/or damage to persons or property as a matter of products liability,
negligence or otherwise, or from any use or operation of any methods, products, instructions, or ideas
contained in the material herein.

Library of Congress Cataloging-in-Publication Data
A catalog record for this book is available from the Library of Congress

British Library Cataloguing-in-Publication Data
A catalogue record for this book is available from the British Library

ISBN: 978-0-12-813557-0

For information on all Academic Press publications visit our
website at https://www.elsevier.com/books-and-journals

Working together
to grow libraries in
developing countries

www.elsevier.com • www.bookaid.org

Publisher: Jonathan Simpson
Acquisition Editor: Glyn Jones
Editorial Project Manager: Natasha Welford
Production Project Manager: Debasish Ghosh
Cover Designer: Mark Rogers

Typeset by SPi Global, India

CONTENTS

CHAPTER 1

Introduction

1.1 AN OVERVIEW OF WIRELESS COMMUNICATIONS

The development of wireless communication has gone through four generations. The first-generation (1G) wireless communication system is analog cellular wireless communication based on analog signal, which mainly provides voice services. It was the booming period for the development of 1G from the 1980s to 1990s. In 1978, Bell labs developed the advanced mobile phone service systems with the operation frequency 900 MHz. It was firstly put into commercial application in Chicago, Illinois, USA. After that, other industrialized countries developed their own cellular wireless communication systems such as total access communication system, Nordic Mobile Telephone and high capacity mobile telecommunication system, etc. The reason why the development of cellular wireless communication systems was so fast is due to: (1) multichannel sharing and frequency reuse technique; (2) complete system functions such as handover, wander, etc.; and (3) availability to civil telephone network. 1G employs the frequency division multiple access (FDMA) technique and voice signals are analogy modulated. Although they achieved great commercial success, some drawbacks that existed in these systems became more and more fatal as the number of users increased, such as low spectral efficiency, limited business lines, low rate data service, low security, high-cost equipment, etc.

To overcome the previous drawbacks of 1G, the second-generation (2G) wireless communication system was proposed and achieved fast development. Compared with 1G, 2G improves the spectral efficiency and supports several business lines (i.e., voices and low rate data services); hence it can be also called narrow-band digital wireless communication systems. Two typical cases, global system for mobile communication (GSM) and interim standard 95 (IS-95) were proposed by Europe and the United States from the middle 1980s, respectively. GSM originated from Europe

OQAM/FBMC for Future Wireless Communications
http://dx.doi.org/10.1016/B978-0-12-813557-0.00001-2

© 2018 Elsevier Ltd.
All rights reserved.

and was designed for the global digital cellular communications. It employs time division multiple access (TDMA) and its data rate achieves 64 Kbps, working on 900 MHz. As the digital cellular standard of North America, IS-95 adopts code division multiple access (CDMA) technique, working on the frequency 900 or 1800 MHz.

Since 1990s, the number of Internet users has increased explosively with the rapid development of Internet. People wanted to access the Internet not only at home and the office, but also from moving locations. Thus, it was an urgent requirement to combine Internet and wireless communication technique. However, 1G and 2G cannot meet the requirement due to the low data rate. To settle the conflict between the huge wireless communication market and limited spectrum resources, the third-generation (3G) wireless communication system was developed. All over the world, three famous standards of 3G are wideband CDMA (WCDMA) in Europe, CDMA 2000 in the United States, and time-division synchronous CDMA (TD-SCDMA) in China. Compared with 1G and 2G, 3G systems employ more frequency bands up to more than 5 MHz and the data rates are 384 Kbps at least and 2 Mbps at most. 3G can transmit either voice or data. Thus, it can provide fast and convenient wireless application such as accessing to the Internet wirelessly. 3G can combine high-speed mobile access and Internet services and achieve the goals of global coverage and seamless connection among different wireless networks.

To meet the requirement of high data rate, the international telecommunications union specified a set of requirements for 4G standards in 2008, that is, the international mobile telecommunications advanced (IMT-Advanced), setting the peak speed to be 100 Mbps for high mobility communications such as in cars and trains. Orthogonal frequency division multiplexing (OFDM) and multi-input multi-output (MIMO) techniques constitute the basis of 4G standards. Compared with CDMA, OFDM has a stronger ability to fight against the frequency selective fading channel, especially for broadband wireless communication systems. The MIMO technique provides higher data rate and better transmission quality by adopting more transmit and receive antennas.

The future wireless communication systems (i.e., fifth-generation [5G] wireless communications) are expected to meet the requirement of higher data rate. For general users, the direct experience of 5G is the very large network speed and its peak may be up to 10 Gbps, which means that the time of downloading a high definition movie may be less than 1 s.

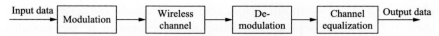

Fig. 1.1 The baseband equivalent block diagram of wireless communication systems.

In addition, with the arising of new wireless Internet applications, future wireless communication systems are surely expected to be massive and tactile interconnected compared to conventional wireless communication systems. On one hand, Cisco predicts that there will be more than 10 billion devices accessing the Internet in 2019. On the other hand, the future Internet will be gradually dominated by control and cognitive Internet applications most of which require tactile latency, such as e-health (less than 10 ms delay), intelligent manufacturing (less than 5 ms delay), and smart transportation (less than 1 ms delay).

However, 5G standards are still at the primary research stage and there are some technique breakthrough to be made for the improvement of spectral efficiency and the reducing of energy consumption and cost. Fig. 1.1 depicts the baseband equivalent block diagram of wireless communication systems. With the growing requirement of data rate, wireless communication systems are employing more and more frequency bands, resulting in very severe effect of multipath fading and the problem of inter-symbol interference (ISI). As is well known, modulations are comprised of multicarrier and single-carrier modulations, and multicarrier modulations are more efficient to overcome the ISI than single-carrier modulations. This book mainly introduces one of novel multicarrier modulations.

The rest of this chapter is organized as follows. Wireless channels are presented in Section 1.2. Section 1.3 introduces single-carrier and multicarrier modulations and specifies their advantages and disadvantages. Then, three types of filter bank multicarrier (FBMC) systems are presented in Section 1.4. In Section 1.5, a comparison between offset quadrature amplitude modulation-based FBMC (OQAM/FBMC) and OFDM is made. Finally, the content and organization of this book are summarized in Section 1.6.

1.2 WIRELESS CHANNELS

Obviously, wireless channels have a great impact on the system performance. Thus, a good understanding of wireless channels lays the foundation for the design of wireless communication systems. Different from the channels of wired communications, wireless channels are very complex, dynamic and time-varying, depending on the environment and mobility of users. User environment can be either cities with high buildings or suburbs with hills, lakes, flat grounds, etc. Users of wireless communications can be summarized into three categories: quasistatic indoor users, low-speed pedestrian users, and high-speed car (train) users. The diversity of environments and the mobility of mobile users make it necessary to study the characteristic of wireless channels.

Wireless communication systems transmit data via electromagnetic waves. Fig. 1.2 shows the typical propagations of electromagnetic waves in wireless communications. Generally, electromagnetic waves in wireless communications can be classified into several categories as follows:

- Direct waves: These are the direct propagation of electromagnetic waves without any obstructions, which are the main propagation modes of ultrashort waves and microwaves. The signal strength via direct waves is the strongest.
- Reflection waves: These are the received signals reflected by buildings or other reflectors. Their signal strengths are weaker than that of direct waves.
- Diffracted waves: These are the electromagnetic waves after diffraction that occurs when electromagnetic waves encounter large buildings, hills, or other objects.
- Scatter waves: These occur after electromagnetic waves pass ground and sky. The signal strength is weak generally.

A defining characteristic of the wireless channels is the variation of the channel strength over time and over frequency. After wireless channels, the

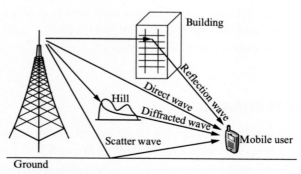

Fig. 1.2 Typical propagations of electromagnetic waves in wireless communications.

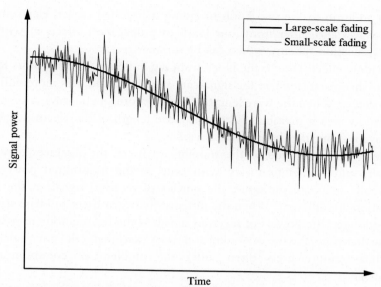

Fig. 1.3 Large-scale fading and small-scale fading.

transmitted signals would suffer propagation losses, which can be roughly divided into two types (i.e., large-scale fading and small-scale fading [1] as shown in Fig. 1.3).

- Large-scale fading: Due to path loss of signal as a function of distance and shadowing by large objects such as buildings and hills. This occurs as the mobile user moves through a distance of the order of the cell size, and is typically frequency independent. From the perspective of statistics, the large-scale fading satisfies the logarithmic normal distribution and its change rate is smaller than the data rate, which is why that it can be also called slow fading.

- Small-scale fading: Due to the constructive and destructive interference of the multiple signal paths between the transmitter and receiver. This occurs at the scale of the order of the carrier wavelength, and is frequency dependent. The small-scale fading satisfies the Rayleigh distribution, the Rice distribution, or the Nakagami distribution generally, whose changing rate is faster than that of large-scale fading. More specifically, the small-scale fading can be classified into space selective fading, frequency selective fading, and time selective fading. The fading characteristics vary over different spaces, frequencies, and times.

In addition to the two kinds of fading mentioned earlier, electromagnetic waves also suffer from four kinds of propagation effects in wireless communication systems, which can be summarized as

- Shadow effect: Due to the block from big buildings, there appears to be one shadow area where the signals are attenuated and thus more difficult to detect. When the wave length of the electromagnetic wave is long, the shadow area is invisible. When the wave length of the electromagnetic wave is short, the shadow area is visible.

- Near-far effect: Due to the mobility of users, the distances between users and base station vary with time. If the transmitted powers of mobile users are the same, the powers of received signals at the base station are different. When the distances between users and base station are longer, the powers of received signals at the base station are weaker. Furthermore, the use of nonlinear devices exacerbates the phenomenon. Thus, remote users often suffer the problem of communication outage.

- Multipath effect: Due to the complex nature of user environment, the received signals consist of direct waves, reflection waves, and scatter waves that have different powers, arrival times, and carrier phases. When these waves superpose together, waves from different paths interfere with each other, which is called the multipath effect. In wireless communication systems, the multipath effect is very common and complex, which results in the very complex receiver. To overcome the multipath effect, multicarrier modulation (MCM) techniques have attracted increasing attention, especially OFDM.

- Doppler effect: The Doppler effect happens when mobile users are moving at a high speed. When the speed is higher, the Doppler effect is more serious, and frequency synchronization has to be performed. In general, we can ignore the Doppler effect if the speed of mobile users is less than 70 km/h.

The channels of wireless communications are completely open, which extremely depend on the propagation distance, landscape, carrier frequency, heights of transmit and receive antennas, etc. Thus, it is very difficult to obtain an accurate channel model of wireless communications. In general, some experimental models are frequently used in projects. Among these models, the Okumura-Hate model is the most efficient and frequently used model in wireless communications. The main conditions for this model are listed as: carrier frequency 150–1500 MHz, propagation distance 1–20 km, antenna height of base station 30–200 m, and antenna height of mobile users 1–10 m. The Okumura-Hate model can be written as

$$L_0 \text{ (dB)} = 69.55 + 26.16 \lg f_c - 13.82 \lg h_b - \alpha(h_m)$$
$$+ (44.9 - 6.55 \lg h_b) \lg d - K_0, \tag{1.1}$$

where L_0 is the averaging path loss, f_c is the carrier frequency, h_b is the antenna height of base station, h_m is the antenna height of mobile users, $\alpha(h_m)$ is the correction factor of mobile antennas, d is the distance between the base station and mobile user, and K_0 is the correction factor of user environment. The average path loss can be different values for different scenarios, which will be discussed in detail as follows.

- For small- and medium-sized cities, it is

$$\alpha(h_m) = [1.1 \lg f_c] h_m - [1.56 \lg f_c - 0.8].$$

- For large cities, it is

$$\alpha(h_m) = 8.29 [\lg 1.54 h_m]^2 - 1.1, \qquad f_c \leq 300 \text{ MHz},$$
$$\alpha(h_m) = 3.2 [\lg 11.75 h_m]^2 - 4.97, \qquad f_c \geq 300 \text{ MHz}.$$

- For urban areas, it is

$$K_0 = 0.$$

- For suburb areas, it is

$$K_0 = 2 [\lg f_c / 28]^2 - 5.4.$$

- For rural areas, it is

$$K_0 = 4.78 [\lg f_c]^2 - 18.33 \lg f_c - 40.98.$$

1.3 SINGLE-CARRIER AND MULTICARRIER MODULATIONS

Currently, wireless communication techniques can be divided into two categories (i.e., single-carrier modulations and multicarrier modulations). Single-carrier modulation systems exploit only one signal frequency to transmit data symbols. Differently, multicarrier modulation systems divide the whole frequency channel into many subcarriers and the high-rate data stream is divided into many low-rate ones transmitted in parallel on subcarriers.

Single-carrier modulation techniques have a long history and have been widely used in many wireless communication systems, including conventional 1G, 2G, 3G wireless communication systems and the uplink of 4G wireless communication systems. Compared with multicarrier modulations, single-carrier modulations have some advantages: (1) the peak to average power ration (PAPR) in single-carrier modulation systems is very low, which is very helpful for the stability of systems and the adoption of low-cost devices in the design of wireless communication systems; (2) compared with multicarrier modulations, single-carrier modulation systems are less sensitive to frequency shift and phase noise, which makes it easier for the time and frequency synchronizations in wireless communication systems, especially for paroxysmal point-to-point communication systems. Due to the previous advantages, single-carrier modulations are still employed in 4G standards, for example, SC-FDMA is the key technique of the uplink in the long-term evolution (LTE) standard. As shown in Fig. 1.4, SC-FDMA employs the N-point discrete Fourier transform (DFT) and M-point ($N < M$) inverse discrete Fourier transform (IDFT) modules at the transmitter, which results in the low PAPR of transmitted signal. Furthermore, the insertion of cyclic prefix (CP) makes it possible to perform channel equalization in the frequency domain so that the simple single-tap equalizer can be employed for channel equalization [2].

However, compared with multicarrier modulations, single-carrier modulations are less capable of dealing with multipath fading channels, resulting in less spectral efficiency. With the development of technology, wireless communication systems are becoming more and more broadband, for example, current TD-LTE occupies 20 MHz broadband and future wireless communication systems will occupy 100 MHz or more. For single-carrier modulation systems, higher bandwidths means smaller symbol interval,

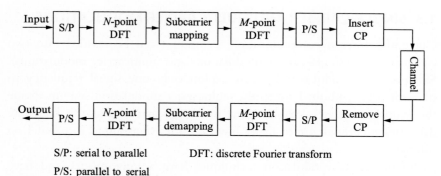

S/P: serial to parallel   DFT: discrete Fourier transform

P/S: parallel to serial

Fig. 1.4 System block of SC-FDMA.

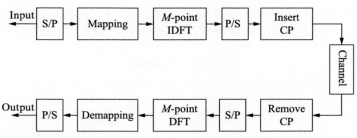

Fig. 1.5 System block of OFDM.

leading to higher sensitivity to multipath fading channels, which makes single-carrier modulation systems vulnerable to ISI. To overcome the ISI, complex multitap equalizers have to be employed, leading to high complexity and cost. Currently, multicarrier modulation systems are attracting more and more attentions due to their ability to fight against multipath fading channel. As the most classic one of multicarrier modulation systems, OFDM was proposed in 1966. However, due to the limitation of hardware devices, OFDM was not put into commercial application until 1980s. So far, OFDM has been widely used in many communication systems, including digital audio broadcasting [3], digital video broadcasting [4], asymmetric digital subscriber line, and wireless local area networks systems [5, 6]. Currently, OFDM has been adopted as the key technique of downlink in the LTE standard [7, 8].

Multicarrier modulation systems divide the whole frequency band into many subcarriers and the high-rate data stream is divided into many low-rate data streams to transmit in parallel on subcarriers. Therefore, the symbol interval in multicarrier modulations is longer than that of single-carrier modulations, which gives multicarrier modulation systems stronger ability against the ISI than single-carrier modulations. Especially for OFDM systems as shown in Fig. 1.5, simple single-tap frequency domain equalizer can be employed with the help of CP, which results in a low-cost receiver. However, multicarrier modulation systems face many technical difficulties that can be summarized as follows:

- *High PAPR.* All multicarrier modulation systems suffer from the problem of high PAPR. The high PAPR is very harmful for the implementation of the transmitter and the whole system performance. The signals of multicarrier modulations are the sum of multiple independent subcarrier signals. When every subcarrier has similar phases, subcarrier signals would be modulated by the signal with similar initial phase, leading to large instantaneous peak power. However, due to the fact that the linear region of the power amplifier is limited, the signals with large power would suffer nonlinear distortion through the amplifier, which results in the serious degradation of system performance. This is why only the downlink of LTE employs multicarrier modulations while the uplink of LTE uses single-carrier modulations.

- *Time and frequency synchronizations.* These are the eternal topics for wireless communications. Since multicarrier signals have longer interval than single-carrier signals, multicarrier modulation systems are less sensitive to time synchronization error compared with single-carrier modulation systems. However, multicarrier modulation systems divide the whole frequency into many subcarriers; hence, each subcarrier occupies a narrower frequency band that is easily affected by frequency shift. Therefore, compared with single-carrier modulations, multicarrier modulation systems are more sensitive to frequency synchronization error. Especially for OFDM systems, frequency shift would cause more serious frequency synchronization error and intercarrier interference (ICI), resulting in the performance degradation.

- *Channel estimation and detection.* Wireless channels have a significant influence on system performance. Therefore, the channel estimation and equalization are very important issues in wireless communications. Channel estimations can be classified into two categories (i.e., blind channel estimations and nonblind channel estimations). Blind channel estimations do not require training sequences and can achieve higher spectral efficiency. However, there exist many drawbacks in blind channel estimations such as bad estimation accuracy, large amount of computation, and poor flexibility. Thus, blind channel estimations are not suitable for real-time systems. Nonblind channel estimations utilize training sequences to achieve channel knowledge and can obtain better estimation accuracy and flexibility, which can be employed in real-time systems. However, use of training sequences occupies time-frequency resources and reduces the spectral efficiency. Thus, how to achieve better performance with less training sequences is very important for wireless communication systems.

- *Multiantenna technique.* In addition to multicarrier modulations, MIMO has also been adopted as one of the key techniques in 4G standards due to its large channel capacity. Therefore, the combination of MIMO and multicarrier modulations will have extensive application prospects in wireless communications.

Despite these technical difficulties, multicarrier modulations have been considered as one of the key techniques in future wireless communication systems. Especially, OFDM has been adopted as the downlink technique of 4G standards. However, OFDM suffers another problem of high spectral sidelobes that is becoming more and more serious due to the fact that spectrum resources are becoming more and more scarce and precious. In cognitive radio systems, spectrum sensing is a very challenging problem as well as data transmission. Since the fast Fourier transform (FFT) operation is usually used for spectral analysis and modulation/demodulation, OFDM has been suggested as a candidate for multicarrier-based cognitive radio systems. However, OFDM has a lot of limitations in the cognitive radio environment. For example, the mutual interference between primary and secondary users in an OFDM-based cognitive radio setting can be reduced only by sacrificing the transmission bandwidth. Furthermore, the channel sensing mechanism requires a dynamic range for the detection of spectrum holes. Unfortunately, the FFT operation in OFDM cannot fulfill this requirement. As a consequence, it is worthwhile to investigate alternative multicarrier processing methods that can overcome the limitations of the FFT operation in OFDM.

1.4 THREE TYPES OF FBMC SYSTEMS

Current physical layer techniques for wireless communications, such as conventional OFDM-based multicarrier communications, are very difficult to support typical applications in massive-connection and tactile Internet. The main reason is that conventional OFDM systems employ time–domain rectangular window with very poor frequency localization, resulting in that communication nodes need to consume much time for signaling overhead and transmission waiting. Thus, conventional OFDM systems are very difficult to meet the requirement of tactile Internet with low latency. In addition, to keep strict orthogonality, it is very hard to adjust system parameters flexibly, such as subcarrier frequency and spacing, which means that conventional OFDM systems are very difficult to be employed in massive-connected environment in future wireless communications with diversified services. Therefore, in academia and industry, it is widely believed that in order to meet the requirement of massive-connection and tactile Internet, the novel multicarrier communication with the abilities of asynchronous transmission and adjustable parameters is becoming one of the most promising research directions for next-generation wireless communications.

Recently, filter bank multicarrier (FBMC) systems have attracted a lot of attention as an alternative scheme to conventional OFDM systems. Due to the employment of the pulse-shaping filter, FBMC systems have lower spectral sidelobes compared with conventional OFDM systems, which means that FBMC have the abilities of asynchronous transmission and adjustable parameters. In addition, as one of multicarrier modulation systems, the symbol interval is extended greatly by dividing the whole frequency band into many subbands, thus FBMC systems are robust to multipath fading channels. Therefore, FBMC systems have been regarded as the potential physical layer techniques in future wireless communication systems.

In this section, three types of FBMC communication systems will be introduced, namely filtered multitone-based FBMC (FMT/FBMC) [9], cosine modulated multitone-based FBMC (CMT/FBMC) [10], and offset quadrature amplitude modulation-based FBMC (OQAM/FBMC) [11, 12]. Compared with OFDM systems, they have lower spectral sidelobes due to the use of the pulse-shaping filters instead of rectangular windows employed in OFDM systems.

1.4.1 FMT/FBMC

FMT/FBMC is an interesting solution for very high-speed digital subscriber line transmission, which is intermediate to other proposed single-carrier and multicarrier methods, as well as providing some unusual advantages in terms of spectrum management, unbundling, and duplexing. Fig. 1.6 presents the structure of an FMT/FBMC multicarrier system.

The complex-valued modulation symbols $x_k(mT)$, $k = 0, 1, \ldots, K - 1$ are obtained from quadrature amplitude modulated (QAM) constellations, where $1/T$ is the symbol rate. After upsampling by a factor of M, each symbol stream is filtered by a baseband filter with frequency characteristic $H(e^{j2\pi f})$ and impulse response $h(t)$. The transmitted signal $s(tT/M)$ is then obtained at the transmission rate of M/T by adding the signals on all K subcarriers. At the receiver, matched filtering (where $\star$ denotes complex conjugation) is employed, followed by downsampling by a factor of M. When $M = K(M > K)$, the filter bank is said to be *critically (noncritically) sampled*.

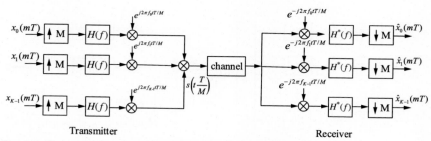

Fig. 1.6 The equivalent baseband diagram of an FMT/FBMC system.

1.4.2 OQAM/FBMC

OQAM/FBMC operates based on QAM symbols whose in-phase and quadrature components are staggered by half the symbol period. In contrast to FMT/FBMC, significant overlap among adjacent bands is allowed in OQAM/FBMC. Successful signal separation is nevertheless possible thanks to a specific signaling arrangement. An introduced orthogonality condition between subcarriers guarantees the received symbols free of ISI and ICI. Carrier orthogonality is achieved through time staggering the in-phase and quadrature components of the subcarrier symbols and designing pulse-shaping filters with good frequency localization property. The equivalent baseband diagram of an OQAM/FBMC system is shown in Fig. 1.7.

The transmitted symbol $x_k(m)$ is real-valued symbol with frequency index k and time index m, and $T/2$ is the interval of real-valued symbols. $x_k(2m)$ and $x_k(2m + 1)$ are obtained by taking the real and imaginary parts of a complex-valued symbol from QAM constellation, respectively. $h(t)$ is a symmetrical real-valued pulse-shaping filter.

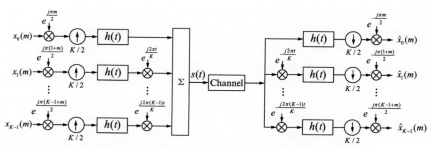

Fig. 1.7 The equivalent baseband diagram of an OQAM/FBMC system.

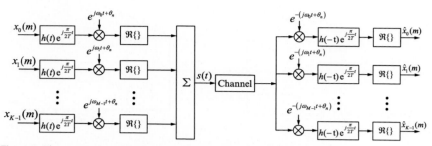

Fig. 1.8 The equivalent baseband diagram of a CMT/FBMC system.

1.4.3 CMT/FBMC

In CMT/FBMC, the subcarrier symbols are pulse amplitude and vestigial side-band (VSB) modulated. Fig. 1.8 presents the structure of a CMT/FBMC multicarrier system. A synthesis filterbank is used to bandlimit a set of PAM symbols to VSB signals and modulate them to various frequency bands.

Fundamentally, VSB filtering is performed through a frequency shifted version of a low-pass filter $h(t)$ centered at $f = \pi/2T$ with impulse response $h(t)e^{j\frac{\pi}{2T}t}$. To extract the kth subcarrier data sequence $x_k(m)$, $s(t)$ is first multiplied by $e^{-j(\omega_k+\theta_k)}$. The received signal is then passed through a low-pass filter whose response is matched to the transmit filter $h(t)e^{j\frac{\pi}{2T}t}$, which is $h(-t)e^{j\frac{\pi}{2T}t}$.

1.4.4 Comparison of Different FBMC Systems

In OQAM/FBMC, each subcarrier band is double sideband modulated and carries a sequence of QAM (i.e., complex-valued) symbols. Opposed to OQAM/FBMC, subcarrier modulation is VSB and the subcarriers carry a sequence of PAM (i.e., real-valued) symbols in CMT/FBMC. Therefore, assuming identical symbol duration and number of subcarriers, the CMT/FBMC signal occupies half the bandwidth of OQAM/FBMC. We denote the data rate of the transmitted signal as R and the corresponding bandwidth as G. It is obvious that the CMT/FBMC signal only provides half of the data rate compared with the OQAM/FBMC signal. FMT/FBMC, on the other hand, introduces guard bands between adjacent subcarriers, which are complex modulated. The width of the guard bands depends on the specific system implementation. Therefore, for an identical number of carriers and identical symbol timing, FMT/FBMC requires more bandwidth than OQAM/FBMC and CMT/FBMC. A comparison of FMT/FBMC, OQAM/FBMC, and CMT/FBMC is shown in Fig. 1.9. Based on the earlier analysis, it is obvious that OQAM/FBMC is superior than FMT/FBMC and

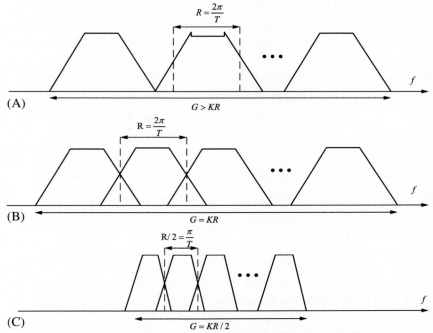

Fig. 1.9 A comparison between FMT/FBMC, OQAM/FBMC, and CMT/FBMC. (A) FMT/FBMC. (B) OQAM/FBMC. (C) CMT/FBMC.

CMT/FBMC in terms of the data rate and utilization efficiency of bandwith. Therefore, we mainly focus on OQAM/FBMC in the rest of this book.

1.5 OQAM/FBMC VS. OFDM

As one of multicarrier modulations, OQAM/FBMC has a long history and its initial concept could be found in 1966 [13]. However, due to the high implementation complexity and the limitation of hardware devices, multicarrier modulation systems, including OFDM and OQAM/FBMC, did not attract much attention and were not put into commercial applications for a long time. From 1980s, with the rapid improvement of hardware, OFDM was firstly put into commercial application and the development of multicarrier modulations entered a boom period. It is more recently that OQAM/FBMC systems have been presented as a viable alternative to the conventional OFDM systems.

OQAM/FBMC is based on a well-designed pulse-shaping filter with good time-frequency localization features. Therefore, the early work about

OQAM/FBMC focuses on the design of pulse-shaping filters and many investigations have been done to achieve good time-frequency localization features. In general, pulse-shaping filters for OQAM/FBMC systems can be summarized as [11]

1. The well-known square root raised cosine (SRRC) function: The continuous-time expression of SRRC functions is given by

$$r_c(t) = \sqrt{F_0} \frac{-4rF_0 \cos(\pi(1+r)F_0 t) + \sin(\pi(1-r)F_0 t)}{(1-(4rF_0 t)^2)\pi F_0 t}, \qquad (1.2)$$

where F_0 is the frequency band and r is the roll-off parameter. The frequency expression of the SRRC function can be obtained as

$$R_c(\nu) = \begin{cases} \frac{1}{\sqrt{F_0}}, & |\nu| \geq (1-r)\frac{F_0}{2}, \\ \frac{1}{\sqrt{F_0}} \cos\left(\frac{\pi}{2r}\left(\frac{|\nu|}{F_0}\right) - \frac{1-r}{2}\right), & (1-r)\frac{F_0}{2} \leq |\nu| \leq (1+r)\frac{F_0}{2}, \\ 0, & |\nu| \geq (1+r)\frac{F_0}{2}. \end{cases}$$
$$(1.3)$$

2. Optimal finite duration pulses (OFDP): The continuous-time expression of OFDP functions can be written as

$$h_c(t) = \sum_{i=0}^{+\infty} \varepsilon_{2i}\varphi_{2i}(t), \qquad (1.4)$$

where $\varphi_{2i}(t)$ is the ith prolate spheroidal wave function and ε_{2i} is real-valued coefficient obtained thanks to an optimization procedure [14].

3. The family of functions called extended Gaussian functions (EGFs).

$$z_{\hat{\alpha},\nu_0,\tau_0}(t) = \frac{1}{2}\left\{\sum_{k=0}^{\infty} d_{k,\hat{\alpha},\nu_0}\left[G_{\hat{\alpha}}\left(t+\frac{k}{\nu_0}\right) + G_{\hat{\alpha}}\left(t-\frac{k}{\nu_0}\right)\right]\right\}$$
$$\times \sum_{l=0}^{\infty} d_{l,1/\hat{\alpha},\tau_0}\cos\left(2\pi l\frac{t}{\tau_0}\right), \qquad (1.5)$$

where $\tau_0\nu_0 = \frac{1}{2}$, $0.528\nu_0^2 \leq \hat{\alpha} \leq 7.568\nu_0$, and $d_{k,\hat{\alpha},\nu_0}$ is a real-valued coefficient and could be computed via the rules described in [15]. $G_{\hat{\alpha}}(t)$ is the Gaussian function as

$$G_{\hat{\alpha}}(t) = (2\hat{\alpha})^{1/4}e^{-\pi\hat{\alpha}t^2}. \qquad (1.6)$$

The above three families of pulse-shaping filters are interesting from different points of view. SRRC functions possess very low spectral sidelobes. OFDP functions have two important properties: (1) OFDP functions do not need to be truncated because they are orthonormal within a finite duration; (2) OFDP functions are optimized to obtain as much energy of pulses as possible within a frequency band. As for the EGF functions, an important property is that they can constitute an orthonormal basis after discretization and be very close to the achievable optimum bound in time-frequency localization. Nevertheless, none of these pulses keep strict orthonormality after both operations of truncation and discretization.

In recent decade, more works on OQAM/FBMC systems have been done. These work mainly focused on the following problems: channel estimation and equalization, PAPR reduction, and combination between OQAM/FBMC and MIMO.

1. *Channel estimation and equalization.* As is well known, wireless channels have a great effect on the system performance. Accurate channel information is necessary to make OQAM/FBMC systems work well. Thus, channel estimation and equalization play an important role in OQAM/FBMC systems. However, different from OFDM systems, the orthogonality condition of OQAM/FBMC systems only holds in the real-number field, which causes the intrinsic imaginary interference to real transmitted symbols at the receiver. Thus, the intrinsic imaginary interference has to be considered for the channel estimation schemes in OQAM/FBMC systems, and the conventional channel estimation methods employed in OFDM systems cannot be used in OQAM/FBMC systems directly. Currently, the channel estimation in OQAM/FBMC systems can be classified into two categories (i.e., frequency-domain channel estimation [16–23] and time-domain channel estimation [24–26]). Similar to OFDM systems, frequency domain channel estimation methods are very simple and single-tap equalizers can be employed for channel equalization. However, frequency domain channel estimation methods work well only when the channel delay spread is small. When the channel delay spread is relatively large, the performances of frequency-domain channel estimation methods degrade seriously. To solve this problem, time-domain channel estimation methods are proposed, which work well even when the channel delay spread is large. As for the equalization, unfortunately, when the channel delay spread is large, simple single-tap equalizers cannot be employed for channel equalization. Therefore, channel estimation and equalization need more investigations in OQAM/FBMC systems.

2. *PAPR reduction.* As one of multicarrier modulations, OQAM/FBMC suffers from the problem of high PAPR. In OFDM systems, symbols are independent with each other. However, in OQAM/FBMC systems,

adjacent symbols are overlapped. Therefore, the PAPR is not only determined by the symbols in certain time, but also their adjacent symbols. The conventional reduction technique of PAPR in OFDM systems maybe not work well in OQAM/FBMC systems. Several PAPR reduction methods have been proposed. In [27], the authors focused on a theoretical and experimental analysis of the PAPR for OQAM/FBMC systems and derived an approximate expression of its complementary cumulative density function. In [28], segmental partial transmit sequence (S-PTS) was proposed, where the overlapped OQAM/FBMC signals are divided into a number of segments, and some disjoint subblocks are divided and multiplied by different phase rotation factors in each segment. In [29], the authors proposed an improved partial transmit sequence (PTS) scheme by employing multiblock joint optimization (MBJO) for the PAPR reduction of OQAM/FBMC signals, called an MBJO-PTS scheme. The MBJO-based scheme exploits the overlapping structure of the OQAM/FBMC signal and jointly optimizes multiple data blocks. However, current PAPR reduction techniques in OQAM/FBMC systems have high complexities. How to achieve good PAPR reduction performance with a low complexity still requires more efforts in future.

3. *Combination between OQAM/FBMC and MIMO.* Due to the fact that the orthogonality condition of OQAM/FBMC systems only holds in the real-number field, the transmitted real-valued symbols would have the intrinsic imaginary interference to each other at the receiver, which results in serious performance degradation for the MIMO with OQAM/FBMC technique (MIMO OQAM/FBMC). Thus, on the contrary with OFDM, it is a challenging work to combine OQAM/FBMC and MIMO. Recently, several solutions to this problem have been proposed. In [30], the authors focused on the transmitter-receiver designs under highly frequency-selective channels in MIMO OQAM/FBMC systems, which can achieve good system performance at the cost of high complexity. In [31], the authors attempted to use the linear and nonlinear transceiver processing techniques to mitigate the effects of ICI and ISI in MIMO OQAM/FBMC systems. In [32], the concept of a two-step receiver combining linear processing and widely linear processing was presented to improve the performance of MIMO OQAM/FBMC systems. In [33], a novel robust beamforming strategy was employed to mitigate the ICI and ISI in MIMO OQAM/FBMC systems under the imperfect CSI conditions. Although these methods can achieve good performance, they suffer from either spectral efficiency loss or high complexity.

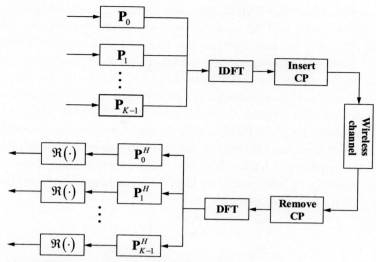

Fig. 1.10 System block of CP-OQAM/FBMC systems.

Recently, several extended versions of OQAM/FBMC have been proposed to enable the simple single-tap equalizer. In [34], the authors combined the single-carrier frequency domain equalization (SC-FDE) techniques with OQAM/FBMC systems (SC-FBMC). In [35], a CP-based OQAM/FBMC system was proposed (denoted as CP-OQAM/FBMC) as shown in Fig. 1.10, where **P** is the filter matrix on each subcarrier and it satisfies the following orthogonality condition

$$\Re\left\{\mathbf{P}_k^H \mathbf{P}_m\right\} = \begin{cases} \mathbf{I}, & k = m \\ \mathbf{0}, & k \neq m, \end{cases} \tag{1.7}$$

where **0** is zero matrix and **I** is identity matrix. Similar to OFDM systems, the CP-OQAM/FBMC system converts a linear convolutional channel to a circular convolutional channel and therefore, makes it possible to employ the simple single-tap equalizer for the channel equalization in the frequency domain. Both of SC-FBMC and CP-OQAM/FBMC exhibit huge simplicity in the aspects of channel equalization and the combination between MIMO and OQAM/FBMC. However, they are sensitive to time-varying channel. In [36], a frequency-spreading FBMC (FS-FBMC) was proposed to improve the ability against multipath fading channels by expanding the length of DFT in the frequency domain. FS-FBMC can simplify the receiver greatly, but it still suffers difficulty for the combination with MIMO.

Currently, OQAM/FBMC is attracting increasing attention. Compared to the conventional OFDM systems, OQAM/FBMC systems have higher spectral efficiency because they do not require CP. Moreover, OQAM/ FBMC systems have lower spectral sidelobes due to the use of the pulse-shaping filters instead of rectangular windows. Recently, the OQAM/ FBMC system has been considered as a promising technique in future wireless communication systems. To understand OQAM/FBMC systems well, we compare the OQAM/FBMC with the most widely used multicarrier modulation technique, OFDM, in the following aspects of orthogonality condition, frequency spectrum, and time-frequency synchronization.

1. *The comparison of orthogonality conditions*. The orthogonality condition of OFDM systems can be written as

$$\sum_{k=-\infty}^{+\infty} e^{j2\pi(m-n)k/K} = \begin{cases} 1, & m = n \\ 0, & m \neq n, \end{cases} \tag{1.8}$$

where m, n are the frequency indexes of subcarriers and K is the subcarrier number in OFDM systems. To satisfy the orthogonality condition, waves of symbols in OFDM systems have to be rectangular windows, which results in higher spectral sidelobes. ICI and ISI may also exist and affect the system performance.

Different from OFDM systems, the orthogonality condition of OQAM/FBMC systems only holds in the real-number field, that is,

$$\Re \left\{ \sum_{k=-\infty}^{+\infty} h\left[k - n\frac{K}{2}\right] h\left[k - q\frac{K}{2}\right] e^{j2\pi(m-p)k/K} e^{j\pi(m+n-p-q)/2} \right\}$$

$$= \begin{cases} 1, & (m, n) = (p, q) \\ 0, & (m, n) \neq (p, q), \end{cases} \tag{1.9}$$

where $h[k]$ is the real-valued pulse-shaping filter with lower sidelobes compared with rectangular windows, which is why OQAM/FBMC systems have lower spectral sidelobes than the conventional OFDM systems. m and p are the frequency indexes of transmitted symbols. n and q are the time indexes of transmitted symbols. K is the subcarrier number of OQAM/FBMC systems. Since the orthogonality condition holds in the real-number field, the transmitted symbols in OQAM/FBMC systems are staggered offset QAM (quadrature amplitude modulation) symbols instead of QAM symbols employed in OFDM systems. More specifically, due to the orthogonality condition in the real-number field, OQAM/

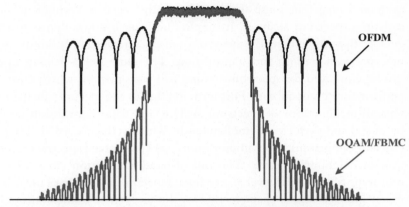

Fig. 1.11 Frequency spectrum comparison between OQAM/FBMC and OFDM.

FBMC systems only transmit real-valued symbols that are from the real and imaginary parts of QAM symbols, respectively. Due to the fact that the real parts and imaginary parts are staggered with half of symbol interval, OQAM/FBMC systems can theoretically achieve the same data rate as OFDM systems.

2. *The comparison of frequency spectrum.* OQAM/FBMC systems exhibit huge advantage over OFDM systems in the aspect of frequency spectrum as shown in Fig. 1.11. That is why OQAM/FBMC systems can be used in future wireless communications. Frequency spectrum is a kind of increasingly scarce and precious resource. However, the rectangular symbols in OFDM systems lead to a $\sin(x)/x$ frequency spectrum with high spectral sidelobes, which reduces the spectrum utilization. Differently, the pulse-shaping filters employed in OQAM/FBMC systems have very low spectral sidelobes. OQAM/FBMC techniques provide a new potential physical layer that offers additional functionalities in the conventional networks and meets the criteria for the existence and acceptance of new concepts, especially for cognitive radio. Cognitive radio is considered as a different context for communication systems. For the physical layer, it implies that two functions have to be performed by the terminals (i.e., the assessment of the spectral environment and opportunistic transmission). The physical layer reports to the upper level decision layers and it receives the instructions for optimal transmission. Pulse-shaping filters in OQAM/FBMC systems can carry out these two functions efficiently and jointly. The following aspects are worth emphasizing: (1) *High-quality*

spectrum sensing. The pulse-shaping filter at the receiver is a high-quality real-time spectrum analyzer. It supplies high-quality raw signals to the devices that implement the spectrum sensing strategies. (2) Simultaneous spectrum sensing and transmission. The same pulse-shaping filter can be employed for spectrum sensing and transmission, which ensures performance compatibility. Furthermore, due to the spectral subchannel separation, functions of spectrum analysis and data transmission could be mixed and performed simultaneously, which is the crucial facility for efficient opportunistic communications. (3) Guaranteed spectral protection of neighboring users. The out-of-band attenuation curve of the prototype filter sets the level of guaranteed spectral protection of the other users. This is a crucial characteristic for the acceptance of the concept of opportunistic communication.

3. *The comparison of time-frequency synchronization.* Time and frequency synchronizations are the eternal topic for wireless communications. For OFDM systems, the rectangular symbols are nonoverlapping to each other in the time domain. However, for OQAM/FBMC systems, the transmitted symbols are overlapping with adjacent symbols. Thus, compared with OFDM systems, OQAM/FBMC systems are more sensitive to the time synchronization error. On the contrary, the rectangular symbols in OFDM systems lead to a high-spectral sidelobe while the transmitted symbols of OQAM/FBMC systems have a very low-spectral sidelobe. Therefore, OFDM systems are more sensitive to the frequency synchronization error compared with OQAM/FBMC systems.

1.6 CONTENTS AND ORGANIZATION

The rest of this book is organized as follows. Chapter 2 introduces the fundamentals of OQAM/FBMC communication systems, including the OQAM/FBMC communication system model, the fast implementation method based on FFT of OQAM/FBMC communication systems, and the CP insertion methods. In Chapter 3, power spectral density of OQAM/FBMC is analyzed. Specifically, the CP insertion methods and spectrum analysis are presented. The prototype filters of OQAM/FBMC are introduced in Chapter 4, where principles of the prototype filter design and filter design methods are provided. In Chapter 5, the tail overhead of OQAM/FBMC signals is discussed and several overhead reduction methods are presented. Chapter 6 is the discussion of PAPR in OQAM/

FBMC systems, where several PAPR reduction methods are introduced. In Chapter 7, effects of time-frequency synchronization are presented as well as pilot design and synchronization recovery. In Chapter 8, channel estimations are introduced for OQAM/FBMC and CP-OQAM/FBMC systems, respectively. Chapter 9 introduces the combination between MIMO and OQAM/FBMC systems and the channel models. Finally, the applications of OQAM/FBMC are introduced in Chapter 10.

REFERENCES

[1] Tse D, Viswanath P. Fundamentals of wireless communication. New York, NY: Cambridge University Press; 2005.
[2] Myung HG, Junsung J, Goodman D. Single carrier FDMA for uplink wireless transmission. IEEE Veh Technol Mag 2007;1(3):30–8.
[3] ETSI EN 300 401 V1.3.3. Radio broadcasting systems; digital audio broadcasting (DAB) to mobile, portable and fixed receiver; 2001.
[4] Nee RV, Prasad R. OFDM for wireless multimedia communications. London, UK: Artech House Publisher; 2000.
[5] IEEE 802.11. Draft supplement to standard for telecommunications and information exchange between systems-LAN/MSN specific requirements. Part 11: wireless MAC and PHY specifications: high speed physical layer in the 5 GHz band; 1999.
[6] IEEE P802.11-TASK GROUP G. Project IEEE 80211g standard for higher rate (20+Mbps) extensions in the 24 GHz band; 2003.
[7] Ghosh A, Ratasuk R, Mondal B, Mangalvedhe N, Thomas T. LTE-advanced: next-generation wireless broadband technology. IEEE Wirel Commun 2010;17(3):10–22.
[8] Deruyck M, Joseph W, Lannoo B, Colle D, Martens L. Designing energy-efficient wireless access networks: LTE and LTE-advanced. IEEE Internet Comput 2013;17(5):39–45.
[9] Cherubini G, Eleftheriou E, Ölçer S. Filtered multitone modulation for very high-speed digital subscriber lines. IEEE J Sel Areas Commun 2002;20(5):1016–28.
[10] Farhang-Boroujeny B, Lin L. Cosine modulated multitone for very high-speed digital subscriber lines. In: IEEE international conference on acoustics, speech, and signal processing (IASSP); 2005.
[11] Siohan P, Siclet C, Lacaille N. Analysis and design of OQAM-OFDM systems based on filterbank theory. IEEE Trans Signal Process 2002;50(5):1170–83.
[12] Fettweis G, Krondorf M, Bittner S. GFDM—generalized frequency division multiplexing. In: IEEE vehicular technology conference (VTC); 2009.
[13] Chang RW. Synthesis of band-limited orthogonal signals for multichannel data transmission. Bell Syst Techn J 1966;1775–96.
[14] Vahlin A, Holte N. Optimal finite duration pulses for OFDM. IEEE Trans Commun 2002;44(1):10–4.
[15] Siohan P, Roche C. Cosine-modulated filterbanks based on extended Gaussian function. IEEE Trans Signal Process 2000;48(11):3052–61.
[16] Lélé C, Legouable R, Siohan P. Channel estimation with scattered pilots in OFDM/OQAM. In: IEEE workshop on signal processing advances in wireless communications (SPAWC); 2008.
[17] Lélé C, Javaudin JP, Legouable R, Skrzypczak A, Siohan P. Channel estimation methods for preamble-based OFDM/OQAM modulations. Eur Trans Telecommun 2007;19(7):741–50.

[18] Du J, Signell S. Novel preamble-based channel estimation for OFDM/OQAM systems. In: IEEE international conference on communication; 2009.

[19] Hu S, Wu G, Li S. Preamble design and iterative channel estimation for OFDM/offset QAM system. J Netw 2009;4(10):963–72.

[20] Katselis D, Kofidis E, Rontogiannis A, Theodoridis S. Preamble-based channel estimation for CP-OFDM and OFDM/OQAM systems: a comparative study. IEEE Trans Signal Process 2010;58(5):2911–6.

[21] Katselis D, Bengtsson M, Rojasa CR, Hjalmarsson H, Kofidis E. On preamble-based channel estimation in OFDM/OQAM systems. In: European signal processing conference; 2011.

[22] Lin H, Siohan P. Robust channel estimation for OFDM/OQAM. IEEE Commun Lett 2009;13(10):724–6.

[23] Kofidis E, Katselis D, Rontogiannis A, Theodoridi S. Preamble-based channel estimation in OFDM/OQAM systems: a review. Signal Process 2013;93(7):2038–54.

[24] Garbo G, Mangione S, Maniscalco V. MUSIC-LS modal channel estimation for an OFDM-OQAM system. In: IEEE international conference on signal processing and communication systems (ICSPCS); 2008.

[25] Haltar LG, Newinger M, Nossek JA. Structured subchannel impulse response estimation for filter bank based multicarrier systems. In: International symposium on wireless communication systems (ISWCS); 2012.

[26] Kong D, Qu D, Jiang T. Time domain channel estimation for OQAM-OFDM systems: algorithms and performance bounds. IEEE Trans Signal Process 2014;63(2):322–30.

[27] Skrzypczak A, Siohan P, Javaudin JP. Analysis of the peak-to-average power ratio for OFDM/OQAM. In: IEEE signal processing advances in wireless communications (SPAWC); 2006.

[28] Ye C, Li Z, Jiang T, Ni C, Qi Q. PAPR reduction of OQAM-OFDM signals using segmental PTS scheme with low complexity. IEEE Trans Broadcast 2014;60(1):141–7.

[29] Qu D, Lu S, Jiang T. Multi-block joint optimization for the peak-to-average power ratio reduction of FBMC-OQAM signals. IEEE Trans Signal Process 2013;61(7): 1605–17.

[30] Caus M, Pérez-Neira AI. Transmitter-receiver designs for highly frequency selective channels in MIMO FBMC systems. IEEE Trans Signal Process 2012;60(12):6519–32.

[31] Soysa M, Rajatheva N, Latva-Aho M. Linear and non-linear transceiver processing for MIMO-FBMC systems. In: IEEE international conference on communications (ICC); 2014.

[32] Cheng Y, Haardt M. Widely linear processing in MIMO FBMC/OQAM systems. In: International symposium on wireless communication systems (ISWCS); 2013.

[33] Payaró M, Pascual-Iserte A, Nájar M. Performance comparison between FBMC and OFDM in MIMO systems under channel uncertainty. In: European wireless conference; 2013.

[34] Lin H, Siohan P. A new transceiver system for the OFDM/OQAM modulation with cyclic prefix. In: IEEE 19th international symposium on personal, indoor and mobile radio communications (PIMRC); 2008.

[35] Gao X, Wang W, Xia XG, Au EKS, You X. Cyclic prefixed OQAM-OFDM and its application to single-carrier FDMA. IEEE Trans Commun 2011;59(5):1467–80.

[36] Berg V, Dore JB, Noguet D. A flexible FS-FBMC receiver for dynamic access in the TVWS. In: IEEE cognitive radio oriented wireless networks and communications (CROWNCOM); 2014.

CHAPTER 2

Fundamentals of OQAM/FBMC

Orthogonal frequency division multiplexing (OFDM) has gained much attention as the most popular multicarrier modulation method, which has been adopted in many wireless standards, such as variations of IEEE 802.11 and IEEE 802.16, 3GPP-LTE, and LTE-Advanced. However, it is noted that OFDM has to face many challenges when considered for adoption in complex networks [1]. Despite its prominence in modern broadband radios, OFDM has some drawbacks for the intended scenario in future. For instance, the use of OFDM in the uplink of multiuser networks, known as orthogonal frequency division multiple access, requires full synchronization of all users' signals at the base station. Such synchronization was found to be very difficult to establish, especially in mobile environments where Doppler shifts are hard to track for different users. Another limitation of OFDM is the unaffordable out-of-band leakage, which makes it hard to coexist with other communication systems.

Recently, the offset quadrature amplitude modulation–based filter bank multicarrier (OQAM/FBMC) has been considered as an alternative approach to OFDM [2–7]. Compared with OFDM systems, FBMC systems have higher spectral efficiency because they do not require cyclic prefix (CP). Moreover, FBMC systems have lower spectral sidelobes due to the use of the pulse-shaping filters instead of a rectangular window. With a well-designed prototype filter, OQAM/FBMC can provide high-spectrum utilization in future key scenarios, such as the machine-type communication, the coordinated multipoint, and the fragmented spectrum. Therefore, OQAM/FBMC is considered as a potential candidate technology for the physical layer of future wireless communications systems [8].

The rest of this chapter is organized as follows. The OQAM/FBMC communication system model is introduced in Section 2.1. In Section 2.2, a fast implementation method based on fast Fourier transform (FFT) of

OQAM/FBMC for Future Wireless Communications
http://dx.doi.org/10.1016/B978-0-12-813557-0.00002-4

© 2018 Elsevier Ltd.
All rights reserved.

OQAM/FBMC communication systems is introduced. To simplify the channel equalization, CP is employed in the OQAM/FBMC communication system in Section 2.3. The summary is made in Section 2.4.

2.1 OQAM/FBMC COMMUNICATION SYSTEM MODEL

Fig. 2.1 presents the system model of the OQAM/FBMC communication system. At the transmitter, the complex input symbols are written as

$$x_k(m) = a_k(m) + jb_k(m), \tag{2.1}$$

where $a_k(m)$ and $b_k(m)$ are the real and imaginary parts of the mth symbol on subcarrier k, respectively. The in-phase and quadrature components are staggered in time domain by $T/2$, where T is the symbol period. The symbols are then passed through a bank of transmission filters and modulated using K subcarrier modulators whose carrier frequencies are $1/T$-spaced apart. The OQAM/FBMC modulated signal is [9]

$$s(t) = \sum_{k=0}^{K-1} \sum_{m=-\infty}^{\infty} \left[a_k(m)h(t - mT) + jb_k(m)h(t - mT - T/2) \right] e^{jk\varphi_t}, \tag{2.2}$$

where $h(t)$ is the impulse response of the prototype filter and $\varphi_t = \frac{2\pi t}{T} + \frac{\pi}{2}$. After that, the OQAM/FBMC modulated signal $s(t)$ is modulated to radio frequency (RF) band and transmitted.

For an ideal transmission system, the received signal at the receiver equals the transmitted signal at the transmitter. After demodulation from RF band, the received signal $r(t)$ is demodulated using K subcarrier demodulators and passed to a bank of matched filters. The filtered signal is then sampled with period T, and the output symbols are

$$\hat{x}_k(m) = \hat{a}_k(m) + j\hat{b}_k(m), \tag{2.3}$$

where $\hat{a}_k(m)$ and $\hat{b}_k(m)$ are the real and imaginary parts of the mth received symbol on subcarrier k, respectively. From [9], we have

$$
\begin{aligned}
\hat{a}_k(m) = \sum_{m'=-\infty}^{\infty} \sum_{k'=0}^{K-1} & \int_{-\infty}^{\infty} h(mT - t) \\
& \times \big\{ a_{k'}(m')h(t - m'T) \cos\left[(k' - k)\varphi_t\right] \\
& - b_{k'}(m')h\left(t - m'T - T/2\right) \sin\left[(k' - k)\varphi_t\right] \big\} \, dt,
\end{aligned} \tag{2.4}
$$

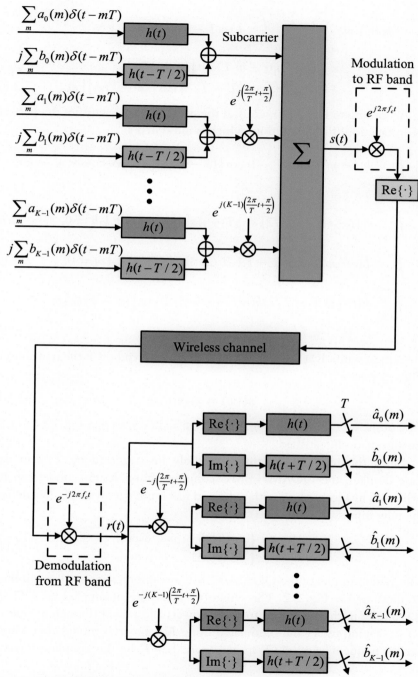

Fig. 2.1 OQAM/FBMC communication system model.

$$\hat{b}_k(m) = \sum_{m'=-\infty}^{\infty} \sum_{k'=0}^{K-1} \int_{-\infty}^{\infty} h\left(mT - t + T/2\right)$$

$$\times \left\{ a_{k'}(m')h(t - m'T) \sin\left[(k' - k)\varphi_t\right] \right.$$
$$\left. + b_{k'}(m')h\left(t - m'T - T/2\right) \cos\left[(k' - k)\varphi_t\right] \right\} dt. \qquad (2.5)$$

According to Eqs. (2.4), (2.5), if the prototype filter satisfies the perfect reconstruction (PR) condition, which is illustrated as

$$\int_{-\infty}^{+\infty} h(t - m'T)h(mT - t) \cos\left[(k' - k)\,\varphi_t\right] dt = \delta(k' - k, m' - m),$$
$$(2.6)$$

$$\int_{-\infty}^{+\infty} h\left(t - m'T - T/2\right) h(mT - t) \sin\left[(k' - k)\,\varphi_t\right] dt = 0, \qquad (2.7)$$

$$\int_{-\infty}^{+\infty} h(t - m'T)h\left(mT - t + T/2\right) \sin\left[(k' - k)\,\varphi_t\right] dt = 0, \qquad (2.8)$$

$$\int_{-\infty}^{+\infty} h(t - m'T - T/2)h(mT - t + T/2) \cos\left[(k' - k)\,\varphi_t\right] dt$$
$$= \delta(k' - k, m' - m), \qquad (2.9)$$

the output symbols at the receiver equal the input symbols at the transmitter. Then, we have

$$\hat{x}_k(m) = x_k(m). \qquad (2.10)$$

Obviously, we can constrain $h(t)$ to be real and even so that Eqs. (2.4), (2.5) are automatically satisfied.

Since the real and imaginary symbols are staggered by $T/2$ in the time domain, the OQAM/FBMC communication system model can be expressed in another way with redefined real input data symbols. We define the new input sequence $d_k(m)$ as

$$d_k(m) = \begin{cases} a_k\left(\frac{m}{2}\right), & m \text{ is even,} \\ b_k\left(\frac{m-1}{2}\right), & m \text{ is odd.} \end{cases} \qquad (2.11)$$

The system model of the OQAM/FBMC communication system can then be illustrated as Fig. 2.2. The symbols $j^m d_k(m)$ are upsampled by $K/2$ (K is typically in the form of 2 to some power) and passed through a bank of pulse-shaping filters and modulated using K subcarrier modulators whose carrier frequencies are $1/T$-spaced apart. Let T_s be the sampling interval

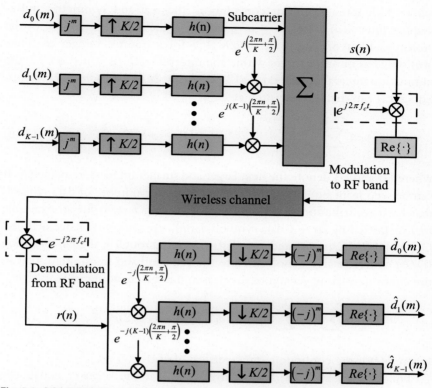

Fig. 2.2 OQAM/FBMC communication system model with real input symbols.

length, then $T_s = T/K$ and the bandwidth is $1/T_s$. The discrete time OQAM/FBMC modulated signal is

$$
\begin{aligned}
s(n) &= \sum_{k=0}^{K-1} \left[h(n) \star \sum_{m=-\infty}^{\infty} j^m d_k(m) \delta(n - mK/2) \right] e^{jk\left(\frac{2\pi n}{K} + \frac{\pi}{2}\right)} \\
&= \sum_{k=0}^{K-1} \sum_{m=-\infty}^{\infty} j^m d_k(m) h(n - mK/2) e^{jk\left(\frac{2\pi n}{K} + \frac{\pi}{2}\right)} \\
&= \sum_{k=0}^{K-1} \sum_{m=-\infty}^{\infty} j^{k+m} d_k(m) h(n - mK/2) e^{j\frac{2\pi kn}{K}}, \quad -\infty < n < \infty,
\end{aligned}
$$

$$(2.12)$$

where $\star$ denotes the linear convolution, the real and symmetric pulse-shaping filter $h(n), -\alpha K/2 \leq n \leq \alpha K/2$, is the impulse response of the prototype filter of length $L_p = \alpha K + 1$ for a positive integer α. Since $h(n)$ only has nonzero values within the interval $-\alpha K/2 \leq n \leq \alpha K/2$, the summation interval of m in Eq. (2.12) is actually $2n/K - \alpha \leq m \leq 2n/K + \alpha$. Eq. (2.12) can then be rewritten as

$$s(n) = \sum_{k=0}^{K-1} \sum_{m=\lceil 2n/K \rceil - \alpha}^{\lfloor 2n/K \rfloor + \alpha} j^{k+m} d_k(m) h(n - mK/2) e^{j\frac{2\pi kn}{K}}, \qquad (2.13)$$

where $\lceil \cdot \rceil$ and $\lfloor \cdot \rfloor$ denote the next larger and smaller integers, respectively. It can be seen from Eq. (2.13) that α determines the number of data symbols that have contribution to a point of $s(n)$. Taking $n = 0$ for an example, $s(0)$ is affected by $2\alpha + 1$ data symbols $d_k(m), -\alpha \leq m \leq \alpha$. After that, the OQAM/FBMC modulated signal $s(n)$ passes through a digital-to-analog (D/A) converter and is modulated to the RF band and transmitted, that is, the continuous-time version of the transmitted baseband signal is

$$s(t) = \sum_{k=0}^{K-1} \sum_{m=\lceil 2t/T \rceil - \alpha}^{\lfloor 2t/T \rfloor + \alpha} j^{k+m} d_k(m) h(t - mT/2) e^{j\frac{2\pi kt}{T}}, \qquad (2.14)$$

where $h(t)$ is the continuous-time version of $h(n)$.

For an ideal channel, the received signal at the receiver equals the transmitted signal at the transmitter. After demodulation from RF band and passing through an analog-to-digital (A/D) converter, the received signal $r(n)$ is demodulated using K subcarrier demodulators and passed to a bank of matched filters. The filtered signal is then downsampled by $K/2$, and the output symbols are

$$\hat{d}_k(m) = \Re \left\{ (-j)^{k+m} \sum_{n=-\infty}^{\infty} r(n) h \left(\frac{mK}{2} - n \right) e^{-j\frac{2\pi kn}{K}} \right\}, \qquad (2.15)$$

where $\Re$ stands for the real part. Similar to the derivation of Eq. (2.13), Eq. (2.15) can be rewritten as

$$\hat{d}_k(m) = \Re \left\{ (-j)^{k+m} \sum_{n=\frac{m-\alpha}{2}K}^{\frac{m+\alpha}{2}K} r(n) h \left(\frac{mK}{2} - n \right) e^{-j\frac{2\pi kn}{K}} \right\}. \qquad (2.16)$$

The recovered QAM symbols $\hat{x}_k(m) = \hat{a}_k(m) + j\hat{b}_k(m)$ are

$$\hat{d}_k(m) = \begin{cases} \hat{a}_k(\frac{m}{2}), & m \text{ is even}, \\ \hat{b}_k(\frac{m-1}{2}), & m \text{ is odd}. \end{cases} \tag{2.17}$$

By substituting Eq. (2.12) into Eq. (2.15), we have

$$\hat{a}_k(m) = \sum_{m'=-\infty}^{\infty} \sum_{k'=0}^{K-1} \sum_{n=-\infty}^{\infty} h(mK - n)$$
$$\times \left\{ a_{k'}(m')h(n - m'K) \cos\left[(k' - k)\left(\frac{2\pi n}{K} + \frac{\pi}{2} \right) \right] \right.$$
$$\left. - b_{k'}(m')h\left(n - m'K - \frac{K}{2} \right) \sin\left[(k' - k)\left(\frac{2\pi n}{K} + \frac{\pi}{2} \right) \right] \right\}, \tag{2.18}$$

$$\hat{b}_k(m) = \sum_{m'=-\infty}^{\infty} \sum_{k'=0}^{K-1} \sum_{n=-\infty}^{\infty} h(mK - n + K/2)$$
$$\times \left\{ a_{k'}(m')h(n - m'K) \sin\left[(k' - k)\left(\frac{2\pi n}{K} + \frac{\pi}{2} \right) \right] \right.$$
$$\left. + b_{k'}(m')h\left(n - m'K - K/2 \right) \cos\left[(k' - k)\left(\frac{2\pi n}{K} + \frac{\pi}{2} \right) \right] \right\}. \tag{2.19}$$

Obviously, Eqs. (2.18), (2.19) are consistent with Eqs. (2.4), (2.5), which indicates that the PR condition is not affected by newly defined real input symbols.

Let $H(z)$ be the z-transform of the filter $h(n)$ and $H_k(z)$, $0 < k \leq K-1$ be the Type 1 polyphase components of $H(z)$, the PR condition in the z domain can be expressed as [10]

$$\tilde{H}_k(z)H_k(z) + \tilde{H}_{k+K/2}(z)H_{k+K/2}(z) = \beta, \quad 0 \leq k \leq K/2 - 1, \tag{2.20}$$

where $\beta > 0$ is a constant, the tilde operation is defined by $\tilde{H}(z) = H^{\star}(1/z^{\star})$ and * denotes the complex conjugate [11]. When the prototype filter satisfies the PR condition in Eq. (2.20), the output symbols at the receiver equal the input symbols at the transmitter. Eq. (2.20) is also the PR condition for cosine modulated filter bank communication systems [11].

2.2 FFT IMPLEMENTATION OF OQAM/FBMC COMMUNICATION SYSTEMS

The conventional implementation structures of OQAM/FBMC communication systems have very high computational complexity. In this section, we present a fast implementation method based on the inverse fast Fourier transform/fast Fourier transform (IFFT/FFT) structure.

The signal on the kth subcarrier can be written as

$$s_k(n) = \sum_{m=0}^{L_p-1} a_k(m)h(n - mK) + jb_k(m)h\left(n - mK - \frac{K}{2}\right). \qquad (2.21)$$

The transmitted signal of OQAM/FBMC systems can then be written as

$$s(n) = \sum_{k=0}^{K-1} s_k(n)e^{jk(\frac{2\pi n}{K} + \frac{\pi}{2})}. \qquad (2.22)$$

For simplicity, let $D_{2l}^k = a_k(l)$ and $D_{2l+1}^k = b_k(l)$. Therefore, the transmission data matrix can be represented as $\mathbf{D} = [\mathbf{D}_0, \mathbf{D}_1, \ldots, \mathbf{D}_{2L-1}]$, where $\mathbf{D}_l = \left[D_l^0, D_l^1, \ldots, D_l^{K-1}\right]^T$, $l = 0, 1, \ldots, 2L - 1$. Eqs. (2.21), (2.22) can then be rewritten as

$$s(n) = \sum_{l=0}^{2L-1} s_l(n), \quad n = 0, 1, \ldots, \left(\alpha + L - \frac{1}{2}\right)K, \qquad (2.23)$$

where

$$s_l(n) = \sum_{k=0}^{K-1} D_l^k h\left(n - \frac{lK}{2}\right) e^{jk(\frac{2\pi n}{K} + \frac{\pi}{2})} e^{j\frac{\pi}{2} \bmod (l,2)}, \qquad (2.24)$$

where $\bmod(\cdot)$ denotes modules operation.

As shown in Fig. 2.3, the lengths of $s_l(n)$ and $s(n)$ are $\alpha K + 1$ and $(M + L - \frac{1}{2})K + 1$, respectively. We have

$$s_l\left(nK + i + \frac{lK}{2}\right) = d_l(i)h(nK + i), \qquad (2.25)$$

where

$$d_l(i) = \frac{1}{K} \sum_{k=0}^{K-1} E_l^k e^{j\frac{2\pi ki}{K}}, \quad E_l^k = KD_l^k e^{j\frac{\pi}{2}(k(2l+1)+ \bmod (l,2))}. \qquad (2.26)$$

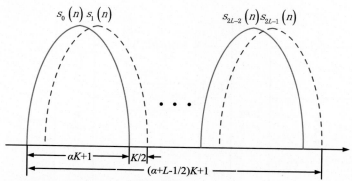

Fig. 2.3 Signal length of OQAM/FBMC systems.

From Eq. (2.25), we have

$$s_l(n) = \sum_{k=0}^{K-1} d_l(n) h_i \left(n - \frac{lK}{2} \right),$$

(2.27)

where

$$h_i(n) = \begin{cases} h(n), & n = \alpha k + i \\ h(n), & n = \alpha K \\ 0, & \text{others} \end{cases}$$

(2.28)

From Eqs. (2.26), (2.27) and (2.28), transmitters of OQAM/FBMC systems consist of three parts. Firstly, the signal E_l^k is obtained by pre-modulation on the transmitted data D_l^k, followed by modulation on the kth subcarrier, which can be implemented by IFFT operation. Then, each output signal is multiplied by the filter $h_i(n - \frac{lK}{2})$. Finally, the sum of signals on K subcarriers is the transmitted signal of OQAM/FBMC systems. The corresponding implementation structure is shown in Fig. 2.4.

At the receiver, the output signal is written as

$$\hat{D}_l^k = \Re \left\{ \sum_{n=0}^{(\alpha+L-\frac{1}{2})K} r(n) h \left(n - \frac{lK}{2} \right) e^{-jk(\frac{2\pi n}{K} + \frac{\pi}{2})} \right\}.$$

(2.29)

For simplicity, we define

$$b(n) = \begin{cases} r(n) h(n - \frac{lK}{2}), & 0 \le n \le (\alpha + L - \frac{1}{2})K \\ 0, & (M + L - \frac{1}{2})K + 1 \le n \le (\alpha + L)K - 1. \end{cases}$$

(2.30)

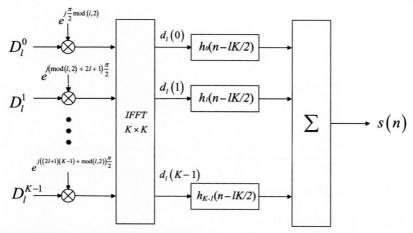

Fig. 2.4 IFFT-based implementation of the transmitter in OQAM/FBMC systems.

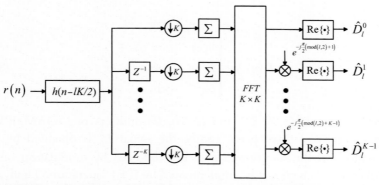

Fig. 2.5 FFT-based implementation of the receiver in OQAM/FBMC systems.

Let $n = mK + i$, where $m = 0, 1, \ldots, \alpha + L - 1, i = 0, 1, \ldots, K - 1$. Eq. (2.29) can then be written as

$$\hat{D}_l^n = \Re \left\{ \sum_{n=0}^{(\alpha+L)K-1} b(n)e^{-jk\left(\frac{2\pi n}{K} + \frac{\pi}{2}\right)} \right\}$$

$$= \Re \left\{ e^{-jk\frac{\pi}{2}} \sum_{i=0}^{K-1} \sum_{m=0}^{\alpha+L-1} b(mK + i)e^{-j\frac{2\pi ki}{K}} \right\}. \quad (2.31)$$

Fig. 2.5 depicts the FFT-based implementation of the receiver in OQAM/FBMC systems. First, multiply the received signal $r(n)$ by the filter $h(n - \frac{lK}{2})$. Then, shift the output signal with i samples and perform K times down–sampling. Then, perform FFT operation on the output signal, followed by the operation of taking real part. Compared with the conventional structure, the FFT-based implementation structure greatly reduces the computational complexity.

2.3 CP-BASED OQAM/FBMC COMMUNICATION SYSTEMS

As is well known, the guard time and CP are employed in OFDM systems generally to deal with multipath fading channels and simplify the channel equalization. However, the conventional OQAM/FBMC systems cannot make use of CP directly, resulting in high complexity on channel equalization. In this section, we present CP-based OQAM/FBMC (CP-OQAM/FBMC) systems, where the linear convolution is replaced by the circular convolution and the time domain channel can be converted into frequency domain parallel channels when the length of CP is larger than the channel delay spread.

We consider an equivalent baseband version of the CP-OQAM/FBMC system with K subcarriers. The transmitted symbols are grouped into blocks with size $2N_d$. Through the circular convolution between the data symbols and the filter, the transmitted signal of one block before the insertion of CP can be written as [7, 12]

$$s(n) = \sum_{k=0}^{K-1} \sum_{m=0}^{2N_d-1} j^{m+k} e^{j\pi mk} a_k(m) h_k(((n - mK/2))_{N_c}), \qquad (2.32)$$

where $N_c = KN_d$, $0 \leq n \leq N_c - 1$, $((\cdot))_{N_c}$ denotes the mod N_c operation, and $a_k(m)$ are real transmitted symbols from the real and imaginary parts of complex symbols in a quadrature amplitude modulation (QAM) constellation. The filter $h_k(n)$ on the $(k + 1)$th subcarrier is

$$h_k(n) = \begin{cases} f_k(n), & 0 \leq n \leq \frac{\alpha K}{2} \\ f_k(n - N_c + 1), & N_c - 1 - \frac{\alpha K}{2} \leq n < N_c - 1 \\ 0, & \text{else}, \end{cases} \qquad (2.33)$$

where

$$f_k(n) = h(n) e^{j2\pi kn/K}, \qquad -\frac{\alpha K}{2} \leq n \leq \frac{\alpha K}{2}, \qquad (2.34)$$

where $h(n)$ is a symmetrical real-valued pulse-shaping filter and satisfies the PR condition, whose length is $\alpha K + 1$. We rewrite Eq. (2.32) as

$$s(n) = \sum_{k=0}^{K-1} \sum_{m=0}^{2N_d-1} j^k e^{j\pi m(k+0.5)} a_k(m) h_k(((n - mK/2))_{N_c}), \qquad (2.35)$$

which will be used later for a compact vector-matrix representation.

In CP-OQAM/FBMC systems, CP can be utilized to avoid the ISI and ICI as shown in Fig. 2.6. The last N_g signal samples of $s(n)$ are inserted prior to the block before transmission, and it is required that $N_g \geq L_g - 1$. After inserting CP, the transmitted signal can be obtained as

$$s_g(n) = \begin{cases} s(N_c - N_g + n), & 0 \leq n \leq N_g - 1, \\ s(n - N_g), & N_g \leq n \leq N_c + N_g - 1. \end{cases} \qquad (2.36)$$

Suppose a multipath fading channel $g(n)$, $n = 0, 1, \ldots, L_g - 1$ interfered by the complex additive white Gaussian noise $\eta(n)$ with zero mean and variance σ^2. The received signal $r_g(n)$ can then be obtained as

$$r_g(n) = s_g(n) \star g(n) + \eta(n), \qquad (2.37)$$

where $\star$ stands for the linear convolution. After removing the CP, the received signal is written as

$$r(n) = r_g(n + N_g), \quad 0 \leq n \leq N_c - 1. \qquad (2.38)$$

The demodulated symbol of the received signal can then be obtained as

$$\hat{a}_k(m) = \sum_{n=0}^{N_c-1} (-j)^k e^{-j\pi m(k+0.5)} h_k(((mK/2 - n))_{N_c}) r(n). \qquad (2.39)$$

For simplicity, we can rewrite Eq. (2.35) in matrix form as

$$\mathbf{s} = \sum_{k=0}^{K-1} j^k \mathbf{H}_k \mathbf{M} \mathbf{\Sigma}_k \mathbf{a}_k, \qquad (2.40)$$

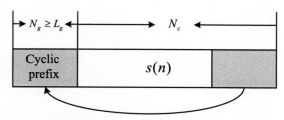

Fig. 2.6 Cyclic prefix in CP-OQAM/FBMC systems.

where $\mathbf{a}_k = [a_k(0), a_k(1), \ldots, a_k(2N_d - 1)]^T$, $\mathbf{s} = [s(0), s(1), \ldots, s(N_c - 1)]^T$, $\boldsymbol{\Sigma}_k = \mathrm{diag}([1, e^{j\pi(k+0.5)}, \ldots, e^{j\pi(2N_d-1)(k+0.5)}])$, $\mathbf{M} = [\mathbf{e}_0, \mathbf{e}_{K/2}, \ldots, \mathbf{e}_{(2N_d-1)K/2}]$, and $\mathbf{H}_k = \mathrm{circ}(\mathbf{h}_k)$ with $\mathbf{h}_k = [h_k(0), h_k(1), \ldots, h_k(N_c - 1)]^T$. Since $\mathbf{H}_k$ is a circular matrix, it can be decomposed as

$$\mathbf{H}_k = \mathbf{W}_{N_c}^H \boldsymbol{\Lambda}_k \mathbf{W}_{N_c}, \tag{2.41}$$

where $\mathbf{W}_{N_c}$ stands for the N_c-point discrete Fourier transform (DFT) matrix with the (m, n)th element $\left[\mathbf{W}_{N_c}\right]_{m,n} = \frac{1}{\sqrt{N_c}} e^{-j2\pi mn/N_c}$, and $\boldsymbol{\Lambda}_k = \mathrm{diag}(\boldsymbol{\lambda}_k)$ with $\boldsymbol{\lambda}_k = \sqrt{N_c} \mathbf{W}_{N_c} \mathbf{h}_k$. Eq. (2.40) can then be rewritten as

$$\mathbf{s} = \mathbf{W}_{N_c}^H \sum_{k=0}^{K-1} \mathbf{P}_k \mathbf{a}_k, \tag{2.42}$$

where $\mathbf{P}_k = j^k \boldsymbol{\Lambda}_k \mathbf{W}_{N_c} \mathbf{M} \boldsymbol{\Sigma}_k$. The orthogonality of CP-OQAM/FBMC systems can be expressed using the matrix form as [12]

$$\Re\left\{\mathbf{P}_k^H \mathbf{P}_m\right\} = \begin{cases} \mathbf{I}_{2N_d}, & k = m \\ \mathbf{0}_{2N_d}, & k \neq m, \end{cases} \tag{2.43}$$

where $\mathbf{0}_{2N_d}$ is a zero matrix with $2N_d \times 2N_d$ dimension. After removing the CP, the received signal is

$$\mathbf{r} = \mathbf{G}_c \mathbf{s} + \boldsymbol{\eta}, \tag{2.44}$$

where $\mathbf{r} = [r(0), r(1), \ldots, r(N_c - 1)]^T$, $\mathbf{G}_c = \mathrm{circ}(\mathbf{g}_c)$ with $\mathbf{g}_c = [g(0), g(1), \ldots, g(L_g-1), 0, 0, \ldots, 0]^T$ as an $N_c \times 1$ vector, and $\boldsymbol{\eta} = \left[\eta(N_g), \eta(N_g + 1), \ldots, \eta(N_g + N_c - 1)\right]^T$ is the channel noise. After N_c-point DFT operation on the received signal $\mathbf{r}$, it can be written as

$$\mathbf{r}_f = \mathbf{W}_{N_c} \mathbf{r} = \mathrm{diag}(\mathbf{G}) \sum_{k=0}^{K-1} \mathbf{P}_k \mathbf{a}_k + \mathbf{W}_{N_c} \boldsymbol{\eta}, \tag{2.45}$$

where $\mathbf{G} = [G(0), G(1), \ldots, G(N_c - 1)]^T$ with $G(m) = \sum_{l=0}^{N_c-1} g(l) e^{-j2\pi ml/N_c}$ as the channel frequency response. Since $\mathrm{diag}(\mathbf{G})$ is a diagonal matrix, the simple single-tap equalizer can be employed when the channel impulse response is known at the receiver. The demodulated symbols are then

$$\hat{\mathbf{a}}_k = \mathbf{P}_k^H [\mathrm{diag}(\mathbf{G})]^{-1} \mathbf{r}_f, \tag{2.46}$$

where $\hat{\mathbf{a}}_k = [\hat{a}_k(0), \hat{a}_k(1), \ldots, \hat{a}_k(2N_d - 1)]^T$. Without the channel noise, it can be obtained as

$$\Re\left\{\hat{\mathbf{a}}_k\right\} = \mathbf{a}_k. \tag{2.47}$$

Under the multipath fading channel, channel estimation and equalization are essential to recover the transmitted symbols perfectly.

2.4 SUMMARY

This chapter introduced the fundamentals of OQAM/FBMC systems. Firstly we introduced the OQAM/FBMC communication system model, where the principles of signal transmit, receive, and reconstruction are presented. Then, we introduced the FFT implementation of OQAM/FBMC systems, which greatly reduces the implementation complexity. Finally, we introduced the CP-based OQAM/FBMC systems, where the time domain channel can be converted into frequency domain parallel channels when the length of CP is larger than the channel delay spread.

REFERENCES

[1] Farhang-Boroujeny B. Filter bank multicarrier modulation: a waveform candidate for 5G and beyond. In: Advances in electrical engineering; 2014.
[2] Farhang-Boroujeny B. OFDM versus filter bank multicarrier. IEEE Signal Process Mag 2011;28(3):92–112.
[3] Bolcskei H. Orthogonal frequency division multiplexing based on offset QAM. In: Advances in Gabor Analysis; 2003, p. 321–52.
[4] Chen D, Qu D, Jiang T. Prototype filter optimization to minimize stopband energy with NPR constraint for filter bank multicarrier modulation systems. IEEE Trans Signal Process 2013;61(1):159–69.
[5] Qu D, Lu S, Jiang T. Multi-block joint optimization for the peak-to-average power ratio reduction of FBMC-OQAM signals. IEEE Trans Signal Process 2013;61(7):1605–17.
[6] Kong D, Qu D, Jiang T. Time domain channel estimation for OQAM-OFDM systems: algorithms and performance bounds. IEEE Trans Signal Process 2014;63(2):322–30.
[7] Kong D, Xia XG, Jiang T, Gao X. Channel estimation in CP-OQAM-OFDM systems. IEEE Trans Signal Process 2014;62(21):5775–86.
[8] Wunder G, et al. 5GNOW: challenging the LTE design paradigms of orthogonality and synchronicity. In: IEEE in vehicular technology conference, Dresden; 2013.
[9] Amini P, Kempter R, Farhang-Boroujeny B. A comparison of alternative filterbank multicarrier methods in cognitive radio systems. In: Software defined radio technical conference, 2006.
[10] Siohan P, Siclet C, Lacaille N. Analysis and design of OQAM-OFDM systems based on filterbank theory. IEEE Trans Signal Process 2002;50(5):1170–83.
[11] Vaidyanathan PP. Multirate systems and filter banks. Englewood Cliffs, NJ: Prentice-Hall; 1993.
[12] Gao X, Wang W, Xia XG, Au EKS, You X. Cyclic prefixed OQAM-OFDM and its application to single-carrier FDMA. IEEE Trans Commun 2011;59(5):1467–80.

CHAPTER 3

Power Spectral Density of OQAM/FBMC

Most of the previous researches are concentrated on offset quadrature amplitude modulation-based filter bank multicarrier (OQAM/FBMC) communication systems without cyclic prefix (CP) insertion, which only works well for narrowband channels and requires complex equalizations at the receiver for broadband channels. In order to simplify the receiver for a broadband channel, similar to the conventional OFDM communication systems, CP can be inserted. Recently, a CP-based OQAM/FBMC communication system was proposed in [1], where the linear convolutions are replaced by circular convolutions and the length of the transmitted signal of each data block is not related to the filter length. The circular structure is also investigated and improved in [2]. Although CP insertions can simplify the receivers and improve the performances for broadband channels, one might think that the spectrum of a CP-based OQAM/FBMC communication system might be much worse than the corresponding OQAM/FBMC communication system without CP insertion. This question will be investigated in this chapter.

In this chapter, we present the system models of CP-OQAM/FBMC, direct CP-OQAM/FBMC (DCP-OQAM/FBMC), and extended CP-based OQAM/FBMC (ECP-OQAM/FBMC) communication systems, investigate the properties of CP-based OQAM/FBMC communication systems, and analyze the power spectral densities (PSDs). We show that the continuous-time transmitted CP-OQAM/FBMC signal after the CP insertion is equivalent to a truncated signal obtained by inserting CP and cyclic suffix (CS) to the discrete-time input data symbols in the OQAM/FBMC communication system, which simplifies the calculation of the PSD of the transmitted CP-OQAM/FBMC signal and also implies that the PSD of the CP-OQAM/FBMC may be affected by the truncation. The ECP-OQAM/FBMC communication system does not truncate the output signal of the OQAM/FBMC with the CP and CS insertions in its input data sequence and has similar spectrum property as the

OQAM/FBMC for Future Wireless Communications
http://dx.doi.org/10.1016/B978-0-12-813557-0.00003-6

© 2018 Elsevier Ltd.
All rights reserved. **39**

OQAM/FBMC communication system. Moreover, we compare three CP-based OQAM/FBMC communication systems with windowed CP-OFDM communication systems in detail. Simulation results show that CP-based OQAM/FBMC communication systems perform better than windowed CP-OFDM communication systems in terms of both the PSD and bit error rate (BER) performances, and the ECP-OQAM/FBMC has the best PSD performance in existing CP-based OQAM/FBMC communication systems with an increase of the data rate overhead.

The rest of the chapter is organized as follows. In Section 3.1, two typical CP-based OQAM/FBMC communication systems, CP-OQAM/FBMC and DCP-OQAM/FBMC, are described. In Section 3.2, one new CP-based OQAM/FBMC communication system, ECP-OQAM/FBMC is introduced. In Section 3.3, the PSD of the transmitted signal in each communication system is analyzed. Simulation results are presented in Section 3.4. Finally, a summary is made in Section 3.5.

3.1 CP INSERTION METHODS

The perfect reconstruction (PR) condition of the OQAM/FBMC communication system ensures the perfect recovery of input data symbols when the channel is ideal. However, for a broadband communication system, the intersymbol interference (ISI) occurs due to the multipath channel. In this case, a complex equalization is needed at the receiver for the OQAM/FBMC communication system, which may not be preferred in wireless communications. To deal with an ISI channel and simplify the equalization at the receiver, CP insertion at the transmitter is an efficient and common method, such as it is used in OFDM communication systems. For the OQAM/FBMC communication system, it can be similarly done. There are two typical CP-based OQAM/FBMC communication systems proposed in the literature. One was proposed in [1], where the linear convolution of an OQAM/FBMC communication system is replaced by the circular convolution, which is called CP-OQAM/FBMC. The other was proposed in [3], where the CP of the whole continuous waveform of an OQAM/FBMC block is inserted, which, for convenience, is called direct CP-OQAM/FBMC (DCP-OQAM/FBMC).

3.1.1 CP-OQAM/FBMC

In order to introduce the CP-OQAM/FBMC proposed in [1], we first describe the system before the insertion of CP, which is in fact a cyclic filter bank [4] and denoted as C-OQAM/FBMC. The CP-OQAM/FBMC is equivalent to inserting a CP to the C-OQAM/FBMC. We first describe the C-OQAM/FBMC, where the transmitted signals are block-based and the linear convolutions in Eqs. (2.12), (2.15) are replaced by circular convolutions. Assume that, for each k, the data symbols $d_k(m)$ are grouped into blocks of size $2N_d$ in terms of m. The input symbols of the lth block on subcarrier k are

$$d_{k,l}(m) = d_k(2lN_d + m), \quad m = 0, 1, \ldots, 2N_d - 1. \tag{3.1}$$

The C-OQAM/FBMC modulated signal and the recovered symbols of the lth block in the ideal channel are

$$s_l(n) = \sum_{k=0}^{K-1} \sum_{m=0}^{2N_d-1} j^{k+m} d_{k,l}(m) h_c\left(\left(\left(n - \frac{mK}{2}\right)\right)_{KN_d}\right) e^{j\frac{2\pi kn}{K}},$$
$$n = 0, 1, \ldots, KN_d - 1, \tag{3.2}$$

$$\hat{d}_{k,l}(m) = \Re\left\{(-j)^{k+m} \sum_{n=0}^{KN_d-1} r_l(n) h_c\left(\left(\left(\frac{mK}{2} - n\right)\right)_{KN_d}\right) e^{-j\frac{2\pi kn}{K}}\right\},$$
$$m = 0, 1, \ldots, 2N_d - 1, \tag{3.3}$$

respectively, where $h_c(((n - x))_y)$ denotes cyclically shifting $h_c(n)$ along n dimension by x positions with period y, $r_l(n)$ denotes the received signal of the lth block and should be equal to $s_l(n)$ in the ideal channel, and $h_c(n)$ of length KN_d is a cyclically shifted version of $h(n)$, which can be expressed as

$$h_c(n) = \begin{cases} h(n), & 0 \leq n \leq \alpha K/2, \\ h(n - KN_d), & KN_d - \alpha K/2 \leq n \leq KN_d - 1, \\ 0, & \text{otherwise,} \end{cases} \tag{3.4}$$

where the length of $h(n)$, L_p and the half data block size N_d should satisfy $L_p = \alpha K + 1 \leq KN_d$. According to Eq. (3.4), it can be seen that $h_c(n)$ (n is nonnegative) is a circularly shifted version of $h(n)$ (n can be either negative or nonnegative), which can be illustrated in Fig. 3.1. Since circular convolutions are used in the C-OQAM/FBMC communication system, the length of the transmitted signal, $s_l(n)$, in the lth block is KN_d, which means that the total length of the transmitted signal of N_b blocks is $N_b K N_d$. Since the total length of the transmitted signal of N_b blocks of input data symbols in the OQAM/FBMC communication system is $N_b K N_d + L_p - 1$,

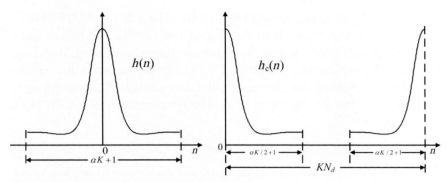

Fig. 3.1 The relationship between $h(n)$ and $h_c(n)$.

it can be seen that the total signal length of C-OQAM/FBMC will not increase, even can reduce after grouping input data symbols.

The continuous–time version of the transmitted baseband C-OQAM/FBMC signal $s_l(n)$ before the CP insertion is

$$s_l^c(t) = \sum_{k=0}^{K-1} \sum_{m=0}^{2N_d-1} j^{k+m} d_{k,l}(m) h_c\left(\left(\left(t - \frac{mT}{2}\right)\right)_{TN_d}\right) e^{j\frac{2\pi kt}{T}}, \quad 0 \le t < TN_d,$$

(3.5)

where $h_c(t)$ is the continuous–time version of $h_c(n)$ and $T = KT_s$. The CP-OQAM/FBMC communication system is to insert CP to the above C-OQAM/FBMC communication system. Assume the CP has N_{cp} sampling points. The time interval length of CP is then $T_{cp} = N_{cp}T_s$. The continuous–time version of the transmitted baseband CP-OQAM/FBMC signal is

$$s_l^{cp}(t) = \begin{cases} s_l^c(t + TN_d - T_{cp}), & 0 \le t < T_{cp}, \\ s_l^c(t - T_{cp}), & T_{cp} \le t < TN_d + T_{cp}. \end{cases}$$

(3.6)

Note that the PR condition for the CP-OQAM/FBMC communication system is not affected by the CP insertion, thus it has the same form as that for the C-OQAM/FBMC communication system. Therefore, we have the following theorem.

Theorem 3.1.1. When the prototype filter $h(n)$ has finite length and the length is not more than KN_d, the PR conditions for the OQAM/FBMC communication system and the CP-OQAM/FBMC communication system are the same.

Proof. As we have explained earlier, we only need to consider the corresponding C-OQAM/FBMC communication system for the CP-OQAM/FBMC communication system. The PR condition for the C-OQAM/FBMC communication system should be the KN_d-point DFT version of the PR condition with the z-transform in Eq. (2.20) for the OQAM/FBMC communication system. According to [4], if we consider the C-OQAM/FBMC communication system in the KN_d-point DFT domain, each point of the transfer function is a sampled version of the transfer function for the OQAM/FBMC communication system on the unit circle in the z-domain. Thus, after Type 1 polyphase decomposition of $h_c(n)$, the PR condition for the C-OQAM/FBMC communication system can be expressed as the N_d-point DFT version of Eq. (2.20). From Eq. (3.4), it is not hard to find that the KN_d-point DFT of $h_c(n)$ and $h(n)$ are the same, which can be expressed as

$$
\begin{aligned}
H_c(p) &= \sum_{n=0}^{KN_d-1} h_c(n) e^{-j\frac{2\pi pn}{KN_d}} \\
&= \sum_{n=0}^{\alpha K/2} h(n) e^{-j\frac{2\pi pn}{KN_d}} + \sum_{n=KN_d-\alpha K/2}^{KN_d-1} h(n-KN_d) e^{-j\frac{2\pi p(n-KN_d)}{KN_d}} \\
&= \sum_{n=0}^{\alpha K/2} h(n) e^{-j\frac{2\pi pn}{KN_d}} + \sum_{n=-\alpha K/2}^{-1} h(n) e^{-j\frac{2\pi pn}{KN_d}} \\
&= \sum_{n=-\alpha K/2}^{\alpha K/2} h(n) e^{-j\frac{2\pi pn}{KN_d}} = H(p).
\end{aligned}
\tag{3.7}
$$

Note that for a length M discrete time signal $x(n)$, its z-transform is equivalent to its M-point DFT. Since the prototype filter $h(n)$ for the OQAM/FBMC communication system has a finite length of not more than KN_d, the z-transform of $h(n)$ is equivalent to the KN_d-point DFT of $h(n)$. The PR condition for the OQAM/FBMC communication system can then also be expressed by the KN_d-point DFT version of Eq. (2.20) in terms of Type 1 polyphase decomposition of $h_c(n)$. As mentioned earlier, the PR condition for the C-OQAM/FBMC is the KN_d-point DFT version of Eq. (2.20) in terms of Type 1 polyphase decomposition of $h_c(n)$.

Therefore, it is clear that the PR conditions for the OQAM/FBMC communication system and the C-OQAM/FBMC communication system are the same.   □

3.1.2 Direct Cyclic Prefixed OQAM/FBMC

The direct CP insertion method was proposed in [3], where a CP is added to each symbol (small block of size K). Since in CP-OQAM/FBMC, a CP is added once to N_d symbols (N_d small blocks). To have a fair comparison, in this section, the direct CP insertion method is applied and the direct cyclic prefixed OQAM/FBMC (DCP-OQAM/FBMC) system is presented, where the OQAM/FBMC input symbols are grouped into blocks of size $2N_d$. The modulated signal of the lth block before the CP insertion is

$$s_l^d(n) = \sum_{k=0}^{K-1} \sum_{m=0}^{2N_d-1} j^{k+m} d_k(m) h\left(n - \frac{\alpha K}{2} - \frac{mK}{2}\right) e^{j\frac{2\pi k(n-\alpha K/2)}{K}},$$

$$n = 0, 1, \ldots, KN_d + \alpha K - K/2. \tag{3.8}$$

The transmitted baseband signal of the lth block after the CP insertion is

$$s_l^{dcp}(t) = \begin{cases} s_l^d(t + TN_d - T_{cp}), & 0 \le t < T_{cp}, \\ s_l^d(t - T_{cp}), & T_{cp} \le t \le TN_d + \alpha T - \frac{T}{2} + T_{cp}, \end{cases} \tag{3.9}$$

where $s_l^d(t)$ is the continuous version of $s_l^d(n)$. We can see that $s_l^{dcp}(t)$ is obtained by directly inserting the tail of $s_l^d(t)$ to the head of $s_l^d(t)$. Since the DCP-OQAM/FBMC communication system uses linear convolutions, the length of the transmitted signal of each block is $TN_d + \alpha T - T/2 + T_{cp}$, which increases the total signal length. Compared with the CP-OQAM/FBMC communication system, the length of the transmitted DCP-OQAM/FBMC signal of each block is $\alpha T - T/2$ longer than that of the CP-OQAM/FBMC communication system.

3.1.3 Equivalence Property of CP-Based OQAM/FBMC

We first rewrite the C-OQAM/FBMC communication system before the CP insertion in terms of $h(n)$ and linear convolution. According to Eqs. (3.2), (3.4), we can obtain

$$s_l(n) = \begin{cases} \sum_{k=0}^{K-1} \sum_{m=\lceil \frac{2n}{K} \rceil - \alpha}^{-1} j^{k+m} d_{k,l}(m + 2N_d) h\left(n - \frac{mK}{2}\right) e^{j\frac{2\pi kn}{K}} \\ \quad + \sum_{k=0}^{K-1} \sum_{m=0}^{\lfloor \frac{2n}{K} \rfloor + \alpha} j^{k+m} d_{k,l}(m) h\left(n - \frac{mK}{2}\right) e^{j\frac{2\pi kn}{K}}, \\ \quad n = 0, 1, \ldots, \frac{(\alpha-1)K}{2}, \\ \sum_{k=0}^{K-1} \sum_{m=\lceil \frac{2n}{K} \rceil - \alpha}^{\lfloor \frac{2n}{K} \rfloor + \alpha} j^{k+m} d_{k,l}(m) h(n - mK/2) e^{j\frac{2\pi kn}{K}}, \\ \quad n = \frac{(\alpha-1)K}{2} + 1, \frac{(\alpha-1)K}{2} + 2, \ldots, KN_d - 1 - \frac{\alpha K}{2}, \\ \sum_{k=0}^{K-1} \sum_{m=\lceil \frac{2n}{K} \rceil - \alpha}^{2N_d - 1} j^{k+m} d_{k,l}(m) h\left(n - \frac{mK}{2}\right) e^{j\frac{2\pi kn}{K}} \\ \quad + \sum_{k=0}^{K-1} \sum_{m=2N_d}^{\lfloor \frac{2n}{K} \rfloor + \alpha} j^{k+m} d_{k,l}(m - 2N_d) h\left(n - \frac{mK}{2}\right) e^{j\frac{2\pi kn}{K}}, \\ \quad n = KN_d - \frac{\alpha K}{2}, KN_d - \frac{\alpha K}{2} + 1, \ldots, KN_d - 1. \end{cases}$$

$$(3.10)$$

To express Eq. (3.10) in a simpler way, we define a new input symbol sequence as

$$d_{k,l}^c(m) = \begin{cases} d_{k,l}(m + 2N_d), & m = -\alpha, -\alpha + 1, \ldots, -1, \\ d_{k,l}(m), & m = 0, 1, \ldots, 2N_d - 1, \\ d_{k,l}(m - 2N_d), & m = 2N_d, 2N_d + 1, \ldots, 2N_d + \alpha - 1. \end{cases}$$

$$(3.11)$$

It can be seen from Eq. (3.11) that $d_{k,l}^c(m)$ is the cyclically extended version of $d_{k,l}(m)$ at both head and tail in terms of m with α symbols each. Eq. (3.10) can then be rewritten as

$$s_l(n) = \sum_{k=0}^{K-1} \sum_{m=\lceil \frac{2n}{K} \rceil - \alpha}^{\lfloor \frac{2n}{K} \rfloor + \alpha} j^{k+m} d_{k,l}^c(m) h\left(n - \frac{mK}{2}\right) e^{j\frac{2\pi kn}{K}},$$

$$n = 0, 1, \ldots, KN_d - 1. \qquad (3.12)$$

Therefore, the transmitted baseband C-OQAM/FBMC signal $s_l(n)$ can be seen as an OQAM/FBMC transmitter with α symbols cyclically inserted to the head and the tail of the input symbols $d_k(m)$. The total length of $s_l(n)$, $-\alpha K/2 \le n \le KN_d + \alpha K/2 + L_p - 2$ obtained by the earlier $d_{k,l}^c(m)$, $-\alpha \le m \le 2N_d - 1 + \alpha$ is $K(N_d + \alpha) + L_p - 1$. However, in Eq. (3.12), we only transmit $s_l(n)$ for $0 \le n \le KN_d - 1$ for the C-OQAM/FBMC communication system and the nonzero head and tail parts are discarded, that is, the C-OQAM/FBMC signal $s_l(n)$, $n = 0, 1, \ldots, KN_d - 1$ is a truncated output of OQAM/FBMC with the cyclically prefixed and suffixed input signal $d_{k,l}^c(m)$ of $d_k(m)$.

Similarly, output symbols in Eq. (3.3) at the C-OQAM/FBMC receiver can be rewritten as

$$\hat{d}_{k,l}(m) = \Re \left\{ (-j)^{k+m} \sum_{n=\frac{m-\alpha}{2}K}^{\frac{m+\alpha}{2}K} r_l^c(n) h\left(\frac{mK}{2} - n\right) e^{-j\frac{2\pi kn}{K}} \right\},$$

$$m = 0, 1, \ldots, 2N_d - 1, \tag{3.13}$$

where $r_l^c(n)$ is obtained by inserting prefix and suffix to the received signal $r_l(n)$, which can be expressed as

$$r_l^c(n) = \begin{cases} r_l(n + KN_d), & n = -\frac{\alpha K}{2}, -\frac{\alpha K}{2} + 1, \ldots, -1, \\ r_l(n), & n = 0, 1, \ldots, KN_d - 1, \\ r_l(n - KN_d), & n = KN_d, KN_d + 1, \ldots, KN_d + \frac{K(\alpha-1)}{2} - 1. \end{cases} \tag{3.14}$$

Note that $r_l^c(n)$ obtained in Eq. (3.14) is purely used for the analysis and is not needed in the receiver. Therefore, the C-OQAM/FBMC demodulated symbols $\hat{d}_{k,l}(m)$ can be obtained by inputting the new signal $r_l^c(n)$ to an OQAM/FBMC receiver. Moreover, we only need to demodulate $2N_d$ symbols of $\hat{d}_{k,l}(m)$ in Eq. (3.13) from the index $m = 0$ to $m = 2N_d - 1$. The implementation of the C-OQAM/FBMC communication system using the OQAM/FBMC communication system is illustrated in Fig. 3.2.

By exploiting the previous relationship between the OQAM/FBMC communication system and the C-OQAM/FBMC communication system, we have the following theorem.

Theorem 3.1.2. *The CP insertion of the continuous-time transmitted baseband signal $s_l^c(t)$ in the CP-OQAM/FBMC communication system is equivalent to a truncated version of the transmitted baseband signal obtained by inserting CP and CS of the discrete-time input symbols $d_{k,l}(m)$ in the OQAM/FBMC communication system.*

Proof. According to Eqs. (3.11), (3.12), the circular convolution can be transformed to linear convolution by periodically extending the input data symbols. Note that the CP insertion can also be seen as a kind of extending the signal. We then define a cyclic prefixed and suffixed version of $d_{k,l}(m)$ as

$$d_{k,l}^{cp}(m) = \begin{cases} d_{k,l}(m + 2N_d), & m = -\left\lceil \frac{2T_{cp}}{T} \right\rceil - \alpha, -\left\lceil \frac{2T_{cp}}{T} \right\rceil - \alpha + 1, \ldots, -1, \\ d_{k,l}(m), & m = 0, 1, \ldots, 2N_d - 1, \\ d_{k,l}(m - 2N_d), & m = 2N_d, 2N_d + 1, \ldots, 2N_d + \alpha - 1. \end{cases} \tag{3.15}$$

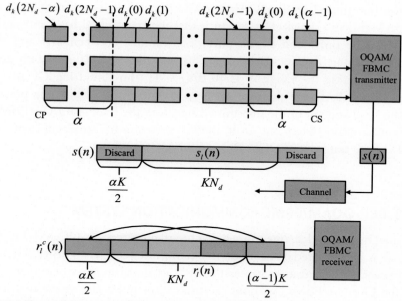

Fig. 3.2 Equivalent implementation of the C-OQAM/FBMC communication system using an OQAM/FBMC communication system.

Similar to Eq. (3.5), it is not hard to see that Eq. (3.6) can be expressed using $d_{k,l}^{cp}(m)$ as

$$s_l^{cp}(t) = \sum_{k=0}^{K-1} \sum_{m=\lceil \frac{2t-2T_{cp}}{T} \rceil - \alpha}^{\lfloor \frac{2t-2T_{cp}}{T} \rfloor + \alpha} j^{k+m} d_{k,l}^{cp}(m) h\left(t - T_{cp} - \frac{mT}{2}\right) e^{j\frac{2\pi k(t-T_{cp})}{T}},$$

$$0 \le t < TN_d + T_{cp}. \tag{3.16}$$

From Eq. (3.16), it can be seen that the signal, $s_l^{cp}(t)$, $0 \le t < TN_d + T_{cp}$ of the CP-OQAM/FBMC communication system can be obtained by truncating the output signal of the OQAM/FBMC communication system with the discrete input sequence $d_{k,l}^{cp}(m)$, where $d_{k,l}^{cp}(m)$ is obtained by inserting CP and CS to the discrete-time input data symbols $d_{k,l}(m)$. □

Different from the CP-OQAM/FBMC communication system, the CP insertion to the continuous time transmitted DCP-OQAM/FBMC signal cannot be obtained by inserting CP or CS to the discrete-time input data symbols, which may complicate the calculation of the PSD of the transmitted DCP-OQAM/FBMC signal and may cause that the spectrum of the transmitted DCP-OQAM/FBMC signal is worse than that of the OQAM/FBMC signal. The reason is that the transmitted baseband signal $s_l^d(t)$ can be only seen as the linear convolution between $h(t)$ and the extended input data symbols of $d_{k,l}(m), 0 \le m \le 2N_d - 1$ with zero symbols outside the range $m \in [0, 2N_d - 1]$, thus, inserting CP or CS to $d_{k,l}(m)$ will change the original head or tail of $s_l^d(t)$. Since the powers of the head and tail of the transmitted signals in the DCP-OQAM/FBMC communication system are small, the CP insertion has small effect on the PSD of the transmitted DCP-OQAM/FBMC signal.

3.2 ECP-OQAM/FBMC COMMUNICATION SYSTEM

From Theorem 3.1.2, one can see that the transmitted CP-OQAM/FBMC signal is a truncated version of the transmitted OQAM/FBMC signal. However, this truncation causes the PSD of signal much worse than that of the OQAM/FBMC without the truncation and is also worse than that of the transmitted DCP-OQAM/FBMC signal. Based on Theorem 3.1.2, a new CP-based OQAM/FBMC communication system was proposed in [5], by adding CP and CS to the input data symbols $d_{k,l}(m)$ of length $2N_d$ of each block to get $d_{k,l}^{cp}(m)$ of length $N_{d,cp} = 2N_d + 2\alpha + \lceil 2T_{cp}/T \rceil$ as shown in Eq. (3.15), then keeping/transmitting the whole signal without any truncation of the output signal of the OQAM/FBMC transmitter with the discrete input sequence $d_{k,l}^{cp}(m)$. Since the above-mentioned output signal is an extended version of the output signal of the CP-OQAM/FBMC, it is named extended CP-OQAM/FBMC (ECP-OQAM/FBMC) and described in detail in the following section.

Assume the data symbols contain N_b blocks. The cyclically prefixed and suffixed data symbols of all blocks are then

$$d_k^{ecp}(m) \overset{\Delta}{=} d_{k,\lfloor m/N_{d,cp} \rfloor}^{cp} \left(\mathrm{mod}(m, N_{d,cp}) - \alpha - \lceil 2T_{cp}/T \rceil \right),$$

$$0 \le m \le N_b N_{d,cp} - 1, \tag{3.17}$$

where $\mathrm{mod}(x, y)$ denotes the remainder of x divided by y. Each block is cyclically prefixed and suffixed as in Eq. (3.15) and the prefix part has $\lceil 2T_{cp}/T \rceil$ more symbols than the suffix part. After the CP and CS insertions, the length of $d_k^{ecp}(m)$ is $N_{d,ecp} = N_b N_{d,cp} = N_b(2N_d + 2\alpha + \lceil 2T_{cp}/T \rceil)$. The transmitted baseband ECP-OQAM/FBMC signal of all N_b blocks is

$$s^{ecp}(t) = \sum_{k=0}^{K-1} h\left(t - \frac{\alpha T}{2}\right) \star \left(\sum_{m=0}^{N_{d,ecp}-1} j^{k+m} d_k^{ecp}(m) \delta\left(t - \frac{mT}{2}\right)\right) e^{j\frac{2\pi k\left(t - \frac{\alpha T}{2}\right)}{T}}$$

$$= \sum_{k=0}^{K-1} \sum_{m=0}^{N_{d,ecp}-1} j^{k+m} d_k^{ecp}(m) h\left(t - \frac{\alpha T}{2} - \frac{mT}{2}\right) e^{j\frac{2\pi k\left(t - \frac{\alpha T}{2}\right)}{T}},$$

$$0 \le t \le N_b\left(TN_d + \alpha T + \left\lceil \frac{2T_{cp}}{T}\right\rceil \frac{T}{2}\right) + \alpha T - T/2. \tag{3.18}$$

From Eq. (3.18), it can be seen that the transmitted baseband signal $s^{ecp}(t)$ on each subcarrier is the whole output of the linear convolution between the upsampled data sequence $j^{m+k} d_k^{ecp}(m)$ and the filter $h(t - \alpha T/2)$ without any truncation of the output signal. Since $h(t)$ only has nonzero values for $-\alpha T/2 \le t \le \alpha T/2$, the summation range of m in Eq. (3.18) is actually $\frac{2t}{T} - 2\alpha \le m \le \frac{2t}{T}$ for each value of $s^{ecp}(t)$ at time t. According to Eqs. (3.15), (3.17), we have

$$s^{ecp}(t) = \sum_{k=0}^{K-1} \sum_{m=0}^{2\alpha + \left\lceil \frac{2T_{cp}}{T}\right\rceil - 1} j^{k+m} d_k^{ecp}(m) h\left(t - \frac{\alpha T}{2} - \frac{mT}{2}\right) e^{j\frac{2\pi k\left(t - \frac{\alpha T}{2}\right)}{T}}$$

$$= \sum_{k=0}^{K-1} \sum_{m=0}^{2\alpha + \left\lceil \frac{2T_{cp}}{T}\right\rceil - 1} j^{k+m} d_k^{ecp}(m + 2N_d) h\left(t - \frac{\alpha T}{2} - \frac{mT}{2}\right) e^{j\frac{2\pi k\left(t - \frac{\alpha T}{2}\right)}{T}}$$

$$= \sum_{k=0}^{K-1} \sum_{m'=2N_d}^{2N_d + 2\alpha + \left\lceil \frac{2T_{cp}}{T}\right\rceil - 1} j^{k+m} d_k^{ecp}(m') h$$

$$\left(t + TN_d - \frac{\alpha T}{2} - \frac{m'T}{2}\right) e^{j\frac{2\pi k\left(t + TN_d - \frac{\alpha T}{2}\right)}{T}}$$

$$= s^{ecp}(t + TN_d), \quad t \in \left[\alpha T, \alpha T + \left\lceil \frac{2T_{cp}}{T}\right\rceil \frac{T}{2}\right). \tag{3.19}$$

We can see from Eq. (3.19) that $s^{ecp}(t), t \in \left[\alpha T, \alpha T + \left\lceil \frac{2T_{cp}}{T}\right\rceil \frac{T}{2}\right)$ is a CP of $s^{ecp}(t), t \in \left[TN_d + \alpha T, TN_d + \alpha T + \left\lceil \frac{2T_{cp}}{T}\right\rceil \frac{T}{2}\right)$ for the block 0. Similarly, the CP range of $s^{ecp}(t)$ of the block l is obtained as $t \in \left[lN_{d,cp}T/2 + \alpha T, lN_{d,cp}T/2 + \alpha T + \left\lceil \frac{2T_{cp}}{T}\right\rceil \frac{T}{2}\right)$.

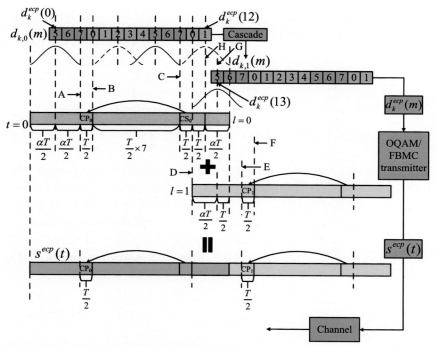

Fig. 3.3 An example of the ECP-OQAM/FBMC transmitter.

In Fig. 3.3, an example of the ECP-OQAM/FBMC transmitter is presented to show that a CP of length T_{cp} is still maintained for each block, where $N_b = 2$, $N_d = 4$, $\alpha = 2$, and $T_{cp} = T/4$. Input data symbols of each block are cyclically prefixed with $\alpha + \left\lceil \frac{2T_{cp}}{T} \right\rceil = 3$ symbols and cyclically suffixed with $\alpha = 2$ symbols. After the linear convolution with $h(t)$ of length αT, each symbol of $d_k^{ecp}(m)$ becomes a pulse of length αT. For each symbol of $d_k^{ecp}(m)$, it only has effect on $s^{ecp}(t)$, $t \in [mT/2, mT/2 + \alpha T]$, for example, $d_k^{ecp}(0)$ only has effect on $s^{ecp}(t)$, $t \in [0, \alpha T]$, which can be illustrated in the figure. For the block 0, $s^{ecp}(t)$, $t \in [\alpha T, \alpha T + T/2)$, which is between the dashed lines A and B in the figure, is only determined by data symbols $d_k^{ecp}(m)$, $m \in [0, 4]$, that is, it can be obtained by inputting sequence $d_{k,0}(5), d_{k,0}(6), d_{k,0}(7), d_{k,0}(0), d_{k,0}(1)$ to an OQAM/FBMC transmitter. $s^{ecp}(t + TN_d)$, $t \in [\alpha T, \alpha T + T/2)$, which is between the dashed lines C and D, is only determined by data symbols $d_k^{ecp}(m)$, $m \in [8, 12]$, that is, can be also obtained by inputting sequence $d_{k,0}(5), d_{k,0}(6), d_{k,0}(7), d_{k,0}(0), d_{k,0}(1)$ to an OQAM/FBMC transmitter due to the cyclic property of $d_k^{ecp}(m)$. It indicates that $s^{ecp}(t) = s^{ecp}(t + TN_d)$ for $t \in [\alpha T, \alpha T + T/2)$, which means the CP range of $s^{ecp}(t)$ for the block 0 is $t \in [\alpha T, \alpha T + T/2)$ that is the CP_0

part between dashed lines A and B shown in Fig. 3.3. Similar to the analysis of the block 0, the CP of $s^{ecp}(t)$ for the block 1 is between the dashed lines E and F. It can be seen that the CP length of each block is actually increased from $T_{cp} = T/4$ to $\lceil 2T_{cp}/T \rceil T/2 = T/2$.

Next, CS_0, which has the same content as CP_0, is not affected by the neighboring block 1, $d_{k,1}^{cp}(m)$, from the right side and therefore CP_0 is a CP for the block 0 at the transmitter. As shown in Fig. 3.3, the most left–hand side symbol, $d_{k,1}(5)$, in the block 1 can only affect up to the part of time length $\alpha T/2$ away from the dashed line G. Note that the α symbols in CS of $d_{k,0}^{cp}(m)$ make that the distance between the dashed lines D and G be $\alpha T/2$, which means that $d_{k,1}(5)$ in the block 1 can only affect $s^{ecp}(t)$ of the block 0 up to the dashed line D. Since the most right–hand side of CS_0, part of the original block 0, $d_{k,0}(m)$, is on the left of the dashed line D, CS_0 is not affected by the neighboring block 1, $d_{k,1}^{cp}(m)$. This also tells the necessity of adding α symbols in CS to the data block 0. For the case of block 1, the most right–hand side symbol, $d_{k,0}(1)$, in the block 0 can only affect up to the part of time length $\alpha T/2$ away from the dashed line H. To guarantee that CP_1 is not affected by $d_{k,0}(1)$ from the left side, the distance between the dashed lines H and E should be greater than $\alpha T/2$. Since the distance between E and F is $\lceil 2T_{cp}/T \rceil T/2$, that is the CP length used to deal with the channel, the distance between the dashed lines H and F should be greater than $\alpha T/2 + \lceil 2T_{cp}/T \rceil T/2$. Note that the $\alpha + \lceil 2T_{cp}/T \rceil$ symbols in CP of $d_{k,1}^{cp}(m)$ make this satisfied. Thus, CP_1 is not affected by the neighboring block 0, $d_{k,0}^{cp}(m)$, from the left side.

Based on the analysis of the previous example, it can be concluded for a general case: The CP of $s^{ecp}(t)$ of each block is not affected by neighboring blocks in $d_k^{ecp}(m)$ as long as there are $\alpha + \lceil 2T_{cp}/T \rceil$ symbols in CP and α symbols in CS in each block of $d_k^{ecp}(m)$. In this case, the CP length in each block in the transmitted ECP-OQAM/FBMC signal is $\lceil 2T_{cp}/T \rceil T/2 \geq T_{cp}$.

The receiver only needs to keep the useful part of the received signal, $t \in \left[lN_{d,cp}T/2 + \alpha T + \left\lceil \frac{2T_{cp}}{T} \right\rceil \frac{T}{2}, lN_{d,cp}T/2 + \alpha T + \left\lceil \frac{2T_{cp}}{T} \right\rceil \frac{T}{2} + TN_d \right)$, of length TN_d of each block and discard the CP and remaining extra overhead parts with no effect on the equalization as long as the CP length is not shorter than the channel length. The useful part of the received signal of each block can then be processed independently the same as the CP-OQAM/FBMC communication system.

Compared with the DCP-OQAM/FBMC communication system, the length of the transmitted ECP-OQAM/FBMC signal of N_b blocks is

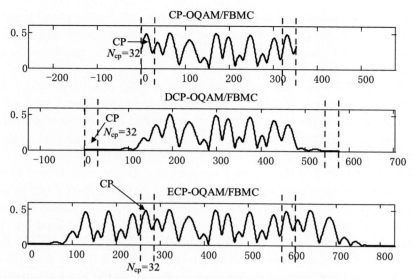

Fig. 3.4 Comparison of the transmitted signals of CP-OQAM/FBMC, DCP-OQAM/FBMC, and ECP-OQAM/FBMC communication systems.

$N_b(\lceil 2T_{cp}/T\rceil T/2 - T_{cp} + T/2) + \alpha T - T/2$ longer than that of the transmitted DCP-OQAM/FBMC signal, which increases the data rate overhead. Examples of transmitted signals of the CP-OQAM/FBMC, DCP-OQAM/FBMC, and ECP-OQAM/FBMC communication systems are compared in Fig. 3.4, where the x-axis denotes the sampling point and y-axis denotes the signal amplitude. The transmitted signals are obtained by $N_b = 1$ block and $N_d = 5$ OQAM/FBMC symbols, the number of subcarriers is $K = 64$, the filter length is $L_p = 4K + 1 = 257$ and the CP length is $N_{cp} = K/2 = 32$. As expected, the length of transmitted DCP-OQAM/FBMC signal has $\alpha K - K/2 = 224$ points more than the transmitted CP-OQAM/FBMC signal and the length of transmitted ECP-OQAM/FBMC signal has $N_b(\lceil 2N_{cp}/K\rceil K/2 - N_{cp} + K/2) + \alpha K - K/2 = 256$ points more than the transmitted DCP-OQAM/FBMC signal. The middle parts of the transmitted signals of three CP-based OQAM/FBMC systems are the same. Since the output signal of an ECP-OQAM/FBMC transmitter can be obtained by inputting $d_k^{ecp}(m)$ to an OQAM/FBMC transmitter without truncation, the PSD of the transmitted ECP-OQAM/FBMC signal is expected to be nearly as good as the transmitted OQAM/FBMC signal.

Remark. The previous CP and CS insertions in $d_{k,l}^{cp}(m)$ of the lth block $d_{k,l}(m)$ for each l can be also viewed as a CP insertion (only) of a shifted version of the original data sequence $d_{k,l}(m)$ with CP length $2\alpha + \lceil 2T_{cp}/T\rceil$: a CP, $d_{k,0}(m)$, $m = 5, 6, 7, 0, 1$, is inserted to $d_{k,0}(m)$, $m = 2, 3, 4, 5, 6, 7, 0, 1$, for the 0th block in the previous example. In this case, the new CP length $2\alpha + \lceil 2T_{cp}/T \rceil$ is the sum of the CP and the CS lengths, and this new CP generates an extra signal of length $\alpha T + \lceil 2T_{cp}/T \rceil T/2$ between any two neighboring blocks when no CP is considered in $s^{ecp}(t)$. If we combine the filter $h(t)$ of length αT and the channel $h_{ch}(t)$ of length $T_{ch} \leq \lceil 2T_{cp}/T \rceil T/2$ as a virtual channel of length $\alpha T + T_{ch}$, which is not longer than the length of the extra signal, then there is no interblock–interference between the original data blocks at the receiver. It seems that this would be simpler than the previous ECP-OQAM/FBMC structure. Although this viewing of the CP only insertion may help to show the necessity of using the total length $2\alpha + \lceil 2T_{cp}/T \rceil$ of both CP and CS in the ECP-OQAM/FBMC communication system, the virtual channel includes the filter $h(t)$ that is used in the OQAM/FBMC communication system for the demodulation and the extra signal may not be just simply deleted at the receiver but some extra caution is needed to produce a similar C-OQAM/FBMC structure as what is done in the CP-OQAM/FBMC communication system. Meanwhile, in the ECP-OQAM/FBMC communication system, the extra signal produced by the CP and the CS can be simply deleted and the remaining part is the same as that in the CP-OQAM/FBMC communication system after deleting its CP at the receiver.

3.3 ANALYSIS IN POWER SPECTRAL DENSITY

In this section, we analyze the PSDs of the transmitted OQAM/FBMC, DCP-OQAM/FBMC, CP-OQAM/FBMC, and ECP-OQAM/FBMC signals.

3.3.1 PSD of the Transmitted OQAM/FBMC Signal

From [6], the PSD of the transmitted OQAM/FBMC signal can be calculated as follows. Define $s(t) = \sum_{k=0}^{K-1} s_k(t)$, where $s_k(t)$ is the transmitted baseband OQAM/FBMC signal on subcarrier k. According to Eq. (2.14), the autocorrelation function, $R_{ss}(t, \tau)$, of the transmitted baseband OQAM/FBMC signal, $s(t)$, is

$$
\begin{aligned}
R_{ss}(t, \tau) &= \mathrm{E}\left[s(t)s^{\star}(t - \tau) \right] \\
&= \mathrm{E}\left[\sum_{k=0}^{K-1} s_k(t) \sum_{k'=0}^{K-1} s_{k'}^{\star}(t - \tau) \right] \\
&= \sum_{k=0}^{K-1}\sum_{k'=0}^{K-1} \mathrm{E}\left[s_k(t)s_{k'}^{\star}(t - \tau) \right].
\end{aligned}
\tag{3.20}
$$

Assume that data symbols $d_k(m)$ are independent and identically distributed. Since $\mathrm{E}\left[d_k(m)d_{k'}^\star(m')\right] = \overline{P}_d\delta(k-k')\delta(m-m')$, $s_k(t) = \sum_{m=-\infty}^{\infty}$ $j^{k+m}d_k(m)h(t-mT/2)e^{j\frac{2\pi kt}{T}}$, where $\overline{P}_d$ is the average power of $d_k(m)$, we have

$$
\mathrm{E}\left[s_k(t)s_{k'}^\star(t-\tau)\right] = \mathrm{E}\left[\sum_{m=-\infty}^{\infty} j^{k+m}d_k(m)h(t-mT/2)e^{j\frac{2\pi kt}{T}}\right.
$$

$$
\times \left.\sum_{m'=-\infty}^{\infty} (-j)^{k'+m'}d_{k'}^\star(m')h^\star(t-\tau-m'T/2)e^{-j\frac{2\pi k'(t-\tau)}{T}}\right]
$$

$$
= \mathrm{E}\left[s_k(t)s_k^\star(t-\tau)\right]\delta(k-k'). \tag{3.21}
$$

Eq. (3.20) can then be rewritten as

$$
R_{ss}(t,\tau) = \sum_{k=0}^{K-1} R_{s_ks_k}(t,\tau)
$$

$$
= \sum_{k=0}^{K-1} \mathrm{E}\left[s_k(t)s_k^\star(t-\tau)\right]
$$

$$
= \sum_{k=0}^{K-1}\sum_{m=-\infty}^{\infty} \mathrm{E}[d_k(m)]^2 h\left(t-\frac{mT}{2}\right)h^\star\left(t-\tau-\frac{mT}{2}\right)e^{j\frac{2\pi kt}{T}}, \tag{3.22}
$$

where $R_{s_ks_k}(t,\tau)$ is the autocorrelation of $s_k(t)$. From Eq. (3.22), we see that $R_{ss}(t,\tau) = R_{ss}(t+T/2,\tau)$ and $R_{s_ks_k}(t,\tau) = R_{s_ks_k}(t+T/2,\tau)$, which means that $s(t)$ and $s_k(t)$ are cyclostationary with period $T/2$. The average autocorrelation function of $s(t)$ is then

$$
\overline{R}_{ss}(\tau) = \sum_{k=0}^{K-1} \overline{R}_{s_ks_k}(\tau)
$$

$$
= \frac{2}{T}\int_0^{T/2}\sum_{k=0}^{K-1}\sum_{m=-\infty}^{\infty} \mathrm{E}[d_k(m)]^2 h\left(t-\frac{mT}{2}\right)h^\star\left(t-\tau-\frac{mT}{2}\right)e^{j\frac{2\pi kt}{T}}dt
$$

$$
= \frac{2\overline{P}_d}{T}\sum_{k=0}^{K-1} e^{j\frac{2\pi k\tau}{T}}\int_{-\infty}^{\infty} h(t)h^\star(t-\tau)dt
$$

$$
= \frac{2\overline{P}_d}{T}\overline{R}_{hh}(\tau)\sum_{k=0}^{K-1} e^{j\frac{2\pi k\tau}{T}}, \tag{3.23}
$$

where $\overline{R}_{hh}(\tau) = \int_{-\infty}^{\infty} h(t)h^{\star}(t-\tau)dt$ is the average autocorrelation function of $h(t)$ and $\overline{R}_{s_k s_k}(\tau) = \frac{2\overline{P}_d}{T}\overline{R}_{hh}(\tau)e^{j\frac{2\pi k\tau}{T}}$ is the average autocorrelation function of $s_k(t)$. Therefore, the PSD of $s(t)$ can be calculated as

$$\begin{aligned}
S_{ss}(w) &= \int_{-\infty}^{\infty} \overline{R}_{ss}(\tau)e^{-jw\tau}d\tau \\
&= \int_{-\infty}^{\infty} \sum_{k=0}^{K-1} \overline{R}_{s_k s_k}(\tau)e^{-jw\tau}d\tau \\
&= \sum_{k=0}^{K-1} S_{s_k s_k}(w) \\
&= \frac{2\overline{P}_d}{T} \sum_{k=0}^{K-1} \int_{-\infty}^{\infty} \overline{R}_{hh}(\tau)e^{-j\left(w-\frac{2\pi k}{T}\right)\tau}d\tau \\
&= \frac{2\overline{P}_d}{T} \sum_{k=0}^{K-1} S_{hh}\left(w - \frac{2\pi k}{T}\right),
\end{aligned} \tag{3.24}$$

where $S_{hh}(w) = \int_{-\infty}^{\infty} \overline{R}_{hh}(\tau)e^{-jw\tau}d\tau$ is the energy spectral density of $h(t)$ and $S_{s_k s_k}(w) = \frac{2\overline{P}_d}{T}S_{hh}(w-\frac{2\pi k}{T})$ is the PSD of $s_k(t)$.

3.3.2 PSD of the Transmitted DCP-OQAM/FBMC Signal

The length of the transmitted baseband DCP-OQAM/FBMC signal, $s_l^d(t)$, of the lth block before the CP insertion is $TN_d + \alpha T - T/2$ and the length of the transmitted baseband DCP-OQAM/FBMC signal, $s_l^{dcp}(t)$, of the lth block after the CP insertion is $TN_d + \alpha T - T/2 + T_{cp}$. It is also not hard to see that $s_l^d(t)$ is cyclostationary with period $TN_d + \alpha T - T/2$ and $s_l^{dcp}(t)$ is cyclostationary with period $TN_d + \alpha T - T/2 + T_{cp}$. Defining the length of DCP-OQAM/FBMC transmitted signal of each block before the CP insertion as $T_{dcp} = TN_d + \alpha T - T/2$, the autocorrelation function, $R_{s_l^{dcp} s_l^{dcp}}(t, \tau)$, of $s_l^{dcp}(t)$ can be obtained from the autocorrelation function, $R_{s_l^d s_l^d}(t, \tau)$, of $s_l^d(t)$ as follows.

1. When $0 \le t < T_{cp}$ and $0 \le t - \tau < T_{cp}$, we have $s_l^{dcp}(t) = s_l^d(t + T_{dcp} - T_{cp})$ and $s_l^{dcp}(t-\tau) = s_l^d(t - \tau + T_{dcp} - T_{cp})$. Then,

$$R_{s_l^{dcp} s_l^{dcp}}(t, \tau) = R_{s_l^d s_l^d}(t + T_{dcp} - T_{cp}, \tau). \tag{3.25}$$

2. When $0 \leq t < T_{cp}$ and $T_{cp} \leq t - \tau < T_{dcp} + T_{cp}$, we have $s_l^{dcp}(t) = s_l^d(t + T_{dcp} - T_{cp})$ and $s_l^{dcp}(t - \tau) = s_l^d(t - \tau - T_{cp})$. Then,

$$R_{s_l^{dcp} s_l^{dcp}}(t, \tau) = R_{s_l^d s_l^d}(t + T_{dcp} - T_{cp}, \tau + T_{dcp}). \qquad (3.26)$$

3. When $T_{cp} \leq t < T_{dcp} + T_{cp}$ and $0 \leq t - \tau < T_{cp}$, we have $s_l^{dcp}(t) = s_l^d(t - T_{cp})$ and $s_l^{dcp}(t - \tau) = s_l^d(t - \tau + T_{dcp} - T_{cp})$. Then,

$$R_{s_l^{dcp} s_l^{dcp}}(t, \tau) = R_{s_l^d s_l^d}(t - T_{cp}, \tau - T_{dcp}). \qquad (3.27)$$

4. When $T_{cp} \leq t < T_{dcp} + T_{cp}$ and $T_{cp} \leq t - \tau \leq T_{dcp} + T_{cp}$, we have $s_l^{dcp}(t) = s_l^d(t - T_{cp})$ and $s_l^{dcp}(t - \tau) = s_l^d(t - \tau - T_{cp})$. Then,

$$R_{s_l^{dcp} s_l^{dcp}}(t, \tau) = R_{s_l^d s_l^d}(t - T_{cp}, \tau). \qquad (3.28)$$

Therefore, the average autocorrelation function of $s_l^{dcp}(t)$ is

$$\overline{R}_{s_l^{dcp} s_l^{dcp}}(\tau) = \begin{cases} \dfrac{\int_0^{T_{dcp}} R_{s_l^d s_l^d}(t,0)dt + \int_{T_{dcp}-T_{cp}}^{T_{dcp}} R_{s_l^d s_l^d}(t,0)dt}{T_{dcp}+T_{cp}}, & \tau = 0, \\[2ex] \dfrac{\int_0^{T_{dcp}-\tau} R_{s_l^d s_l^d}(t,\tau)d\tau + \int_{T_{dcp}-T_{cp}}^{T_{dcp}-\tau} R_{s_l^d s_l^d}(t,\tau)dt}{T_{dcp}+T_{cp}} \\[2ex] \quad + \dfrac{\int_{T_{dcp}-\tau}^{T_{dcp}} R_{s_l^d s_l^d}(t,\tau - T_{dcp})dt}{T_{dcp}+T_{cp}}, & 0 < \tau < T_{cp}, \\[2ex] \dfrac{\int_0^{T_{dcp}-\tau} R_{s_l^d s_l^d}(t,\tau)dt + \int_{T_{dcp}-T_{cp}}^{T_{dcp}} R_{s_l^d s_l^d}(t,\tau - T_{dcp})dt}{T_{dcp}+T_{cp}}, & T_{cp} \leq \tau < T_{dcp}, \\[2ex] \dfrac{\int_{T_{dcp}-T_{cp}}^{2T_{dcp}-\tau} R_{s_l^d s_l^d}(t,\tau - T_{dcp})dt}{T_{dcp}+T_{cp}}, & T_{dcp} \leq \tau < T_{dcp} + T_{cp}, \\[2ex] R_{s_l^d s_l^d}^{\star}(-\tau), & -T_{dcp} - T_{cp} < \tau < 0, \\[1ex] 0, & |\tau| \geq T_{dcp} + T_{cp}, \end{cases} \qquad (3.29)$$

where $R_{s_l^d s_l^d}(t, \tau)$ can be obtained by

$$R_{s_l^d s_l^d}(t, \tau) = \mathrm{E}\left[s_l^d(t) \left(s_l^d(t - \tau) \right)^{\star} \right]$$

$$= \mathrm{E}\left[\sum_{k=0}^{K-1} \sum_{m=0}^{2N_d-1} j^{k+m} d_{k,l}(m) \, h\left(t - \frac{\alpha T}{2} - \frac{mT}{2} \right) e^{j\frac{2\pi k\left(t - \frac{\alpha T}{2}\right)}{T}} \right.$$

$$\times \sum_{k=0}^{K-1} \sum_{m'=0}^{2N_d-1} j^{-k-m'} d_{k,l}(m')$$

$$h^\star\left(t-\tau-\frac{\alpha T}{2}-\frac{mT}{2}\right)e^{-j\frac{2\pi k\left(t-\tau-\frac{\alpha T}{2}\right)}{T}}\Bigg]$$

$$=\overline{P}_d\sum_{k=0}^{K-1}\sum_{m=0}^{2N_d-1}h\left(t-\frac{\alpha T}{2}-\frac{mT}{2}\right)$$

$$h^\star\left(t-\tau-\frac{\alpha T}{2}-\frac{mT}{2}\right)e^{j\frac{2\pi k\tau}{T}}$$

$$=\overline{P}_d\sum_{k=0}^{K-1}\sum_{m=0}^{2N_d-1}R_{hh}\left(t-\frac{\alpha T}{2}-\frac{mT}{2},\tau\right)e^{j\frac{2\pi k\tau}{T}},\qquad(3.30)$$

where $R_{hh}(t,\tau)=h(t)h^\star(t-\tau)$ is the autocorrelation function of $h(t)$. The PSD of $s_l^{dcp}(t)$ is then

$$S_{s_l^{dcp}s_l^{dcp}}(w)=\int_{-\infty}^{\infty}\overline{R}_{s_l^{dcp}s_l^{dcp}}(\tau)e^{-jw\tau}\,d\tau.\qquad(3.31)$$

Since $S_{s_l^{dcp}s_l^{dcp}}(w)$ is not related to the block index l, the PSD of the transmitted baseband ECP-OQAM/FBMC signal, $s^{dcp}(t)$, consisting of N_b blocks is the same as the PSD of each block, that is, $S_{s^{dcp}s^{dcp}}(w)=S_{s_l^{dcp}s_l^{dcp}}(w)$.

3.3.3 PSD of the Transmitted CP-OQAM/FBMC Signal

We can apply the CP equivalence property in Theorem 3.1.2 for CP-OQAM/FBMC and OQAM/FBMC communication systems to estimate the PSD of the transmitted CP-OQAM/FBMC signal in a simpler way than that for the transmitted DCP-OQAM/FBMC signal. Denote the transmitted baseband CP-OQAM/FBMC signal on subcarrier k by $s_{l,k}^{cp}(t)$, which is cyclostationary with the period TN_d+T_{cp}. According to Eq. (3.16), the average autocorrelation function of $s_{l,k}^{cp}(t)$ is

$$\overline{R}_{s_{l,k}^{cp}s_{l,k}^{cp}}(\tau)=\frac{1}{TN_d+T_{cp}}\int_0^{TN_d+T_{cp}}E\left[s_{l,k}^{cp}(t)\left[s_{l,k}^{cp}(t-\tau)\right]^\star\right]dt$$

$$=\frac{1}{TN_d+T_{cp}}\int_{-T_{cp}}^{TN_d}E\Bigg[\sum_{m=\lceil\frac{2t}{T}\rceil-\alpha}^{\lfloor\frac{2t}{T}\rfloor+\alpha}j^{k+m}d_{k,l}^{cp}(m)h\left(t-\frac{mT}{2}\right)e^{j\frac{2\pi kt}{T}}$$

$$\times\sum_{m'=\lceil\frac{2t}{T}\rceil-\alpha}^{\lfloor\frac{2t}{T}\rfloor+\alpha}j^{-k-m'}d_{k,l}^{cp}(m')h^\star\left(t-\tau-\frac{mT}{2}\right)e^{-j\frac{2\pi k(t-\tau)}{T}}\Bigg]dt$$

$$
= \frac{e^{j\frac{2\pi k\tau}{T}}}{TN_d + T_{cp}} \left[\int_{-T_{cp}}^{TN_d} \sum_{m=\lceil \frac{2t}{T} \rceil - \alpha}^{\lfloor \frac{2t}{T} \rfloor + \alpha} \overline{P}_d h\left(t - \frac{mT}{2}\right) \right.
$$

$$
h^*\left(t - \tau - \frac{mT}{2}\right) dt
$$

$$
+ \int_{-T_{cp}}^{TN_d} \sum_{m=-\lceil \frac{2T_{cp}}{T} \rceil - \alpha}^{\alpha - 1} (-1)^{N_d} \overline{P}_d h\left(t - \frac{mT}{2}\right)
$$

$$
h^*\left(t - \tau - \frac{(m + 2N_d)T}{2}\right) dt
$$

$$
+ \int_{-T_{cp}}^{TN_d} \sum_{m=2N_d-\alpha-\lceil \frac{2T_{cp}}{T} \rceil}^{2N_d+\alpha-1} (-1)^{N_d} \overline{P}_d h\left(t - \frac{mT}{2}\right)
$$

$$
\left. h^*\left(t - \tau - \frac{(m - 2N_d)T}{2}\right) dt \right]
$$

$$
= \frac{\overline{P}_d e^{j\frac{2\pi k\tau}{T}}}{TN_d + T_{cp}} \left[\int_{-T_{cp}}^{TN_d} \sum_{m=\lceil \frac{2t}{T} \rceil - \alpha}^{\lfloor \frac{2t}{T} \rfloor + \alpha} R_{hh}\left(t - \frac{mT}{2}, \tau\right) dt \right.
$$

$$
+ \int_{-T_{cp}}^{T\alpha} \sum_{m=-\lceil \frac{2T_{cp}}{T} \rceil - \alpha}^{\alpha - 1} (-1)^{N_d} R_{hh}\left(t - \frac{mT}{2}, \tau + TN_d\right) dt
$$

$$
+ \int_{TN_d-T\alpha-T_{cp}}^{TN_d} \sum_{m=2N_d-\alpha-\lceil \frac{2T_{cp}}{T} \rceil}^{2N_d+\alpha-1}
$$

$$
\left. (-1)^{N_d} R_{hh}\left(t - \frac{mT}{2}, \tau - TN_d\right) dt \right]. \tag{3.32}
$$

The PSD of the transmitted baseband CP-OQAM/FBMC signal $s_{l,k}^{cp}(t)$ is then

$$
S_{s_{l,k}^{cp} s_{l,k}^{cp}}(w) = \int_{-\infty}^{\infty} \overline{R}_{s_{l,k}^{cp} s_{l,k}^{cp}}(\tau) e^{-jw\tau} d\tau. \tag{3.33}
$$

Similar to Eq. (3.24), the PSD of $s_l^{cp}(t)$ is the summation of the PSDs of the transmitted baseband CP-OQAM/FBMC signal on all subcarriers, which can be expressed as

$$S_{s_l^{cp} s_l^{cp}}(w) = \sum_{k=0}^{K-1} S_{s_{l,k}^{cp} s_{l,k}^{cp}}(w). \tag{3.34}$$

Since $S_{s_l^{cp} s_l^{cp}}(w)$ is not related to the block index l, the PSD of the transmitted baseband CP-OQAM/FBMC signal, $s^{cp}(t)$, consisting of N_b blocks is the same as the PSD of each block, that is, $S_{s^{cp} s^{cp}}(w) = S_{s_l^{cp} s_l^{cp}}(w)$.

3.3.4 PSD of the Transmitted ECP-OQAM/FBMC Signal

Since the transmitted baseband ECP-OQAM/FBMC signal is equivalent to the whole output signal obtained by inputting $d_k^{ecp}(m)$ to an OQAM/FBMC transmitter, we can utilize this property to simplify the calculation of the PSD of the baseband signal. Define $c_k(t) = \sum_{m=0}^{N_{d,ecp}-1} j^{k+m} d_k^{ecp}(m)\delta(t - \alpha T/2 - mT/2)e^{-j\pi kL}$ of length $T_{c_k} = T/2(N_{d,ecp} - 1)$ as the continuous-time input data stream for $t \in [0, T_{c_k}]$. The PSD of the transmitted baseband ECP-OQAM/FBMC signal, $s_k^{ecp}(t)$, on subcarrier k, is then calculated as

$$
\begin{aligned}
S_{s_k^{ecp} s_k^{ecp}}(w) &= \mathrm{E}\left[\frac{1}{T_{ecp}}\left| \int_0^{T_{ecp}} s_k^{ecp}(t)e^{-jwt}dt \right|^2 \right] \\
&= \mathrm{E}\left[\frac{1}{T_{ecp}}\left| \int_0^{T_{ecp}} (h(t) \star c_k(t))\, e^{j\frac{2\pi k}{T}} e^{-jwt}dt \right|^2 \right] \\
&= S_{hh}\left(w - \frac{2\pi k}{T}\right) \mathrm{E}\left[\frac{T_{c_k}}{T_{ecp}}\frac{1}{T_{c_k}}\left| \int_0^{T_{c_k}} c_k(t)e^{-j\left(w - \frac{2\pi k}{T}\right)t}dt \right|^2 \right] \\
&= \frac{T_{c_k}}{T_{ecp}} S_{hh}\left(w - \frac{2\pi k}{T}\right) S_{c_k c_k}\left(w - \frac{2\pi k}{T}\right),
\end{aligned}
\tag{3.35}
$$

where $S_{c_k c_k}(w)$ is the PSD of $c_k(t)$ and $T_{ecp} = T/2(N_{d,ecp} - 1) + \alpha T$ is the length of $s^{ecp}(t)$. The average autocorrelation function of $c_k(t)$ is

$$
\begin{aligned}
\overline{R}_{c_k c_k}(\tau) &= \frac{1}{T_{c_k}} \int_0^{T_{c_k}} c_k(t)c_k^{\star}(t - \tau)dt \\
&= \frac{1}{T_{c_k}} \int_0^{T_{c_k}} \mathrm{E}\left[\sum_{m=0}^{N_{d,ecp}-1} j^{k+m} d_k^{ecp}(m)\delta\left(t - \frac{\alpha T}{2} - \frac{mT}{2}\right) e^{-j\pi k\alpha} \right.
\end{aligned}
$$

$$\times \sum_{m=0}^{N_{d,ecp}-1} j^{-k-m} \left(d_k^{ecp}(m)\right)^{\star} \delta\left(t - \tau - \frac{\alpha T}{2} - \frac{mT}{2}\right) e^{j\pi k\alpha} \Bigg] dt$$

$$= \frac{\delta(\tau)}{T_{c_k}} \overline{P}_d N_{d,ecp} + \frac{\delta(\tau + TN_d)}{T_{c_k}} (-1)^{N_b} \overline{P}_d N_b \left(2\alpha + \left\lceil \frac{2T_{cp}}{T} \right\rceil\right)$$

$$+ \frac{\delta(\tau - TN_d)}{T_{c_k}} (-1)^{N_b} \overline{P}_d N_b \left(2\alpha + \left\lceil \frac{2T_{cp}}{T} \right\rceil\right). \tag{3.36}$$

Thus, the PSD of $c_k(t)$ can be calculated as

$$S_{c_k c_k}(w) = \int_{-\infty}^{\infty} \overline{R}_{c_k c_k}(\tau) e^{-jw\tau} d\tau = \frac{N_{d,ecp} \overline{P}_d}{T_{c_k}}$$

$$+ \frac{(-1)^{N_d} N_b \left(2\alpha + \lceil 2T_{cp}/T \rceil\right) \overline{P}_d}{T_{c_k}} 2\cos(TN_d w). \tag{3.37}$$

By substituting Eq. (3.37) into Eq. (3.35), Eq. (3.35) can be rewritten as

$$S_{s_k^{ecp} s_k^{ecp}}(w) = \frac{T_{c_k}}{T_{ecp}} S_{hh}\left(w - \frac{2\pi k}{T}\right) S_{c_k c_k}\left(w - \frac{2\pi k}{T}\right)$$

$$= \frac{N_{d,ecp}}{T_{ecp}} \overline{P}_d S_{hh}\left(w - \frac{2\pi k}{T}\right)$$

$$+ \frac{(-1)^{N_d} 2 N_b \left(2\alpha + \lceil 2T_{cp}/T \rceil\right)}{T_{ecp}} \overline{P}_d S_{hh}\left(w - \frac{2\pi k}{T}\right) \cos(TN_d w)$$

$$= \frac{TN_{d,ecp}}{2T_{ecp}} S_{s_k s_k}(w)$$

$$+ \frac{(-1)^{N_d} T N_b \left(2\alpha + \lceil 2T_{cp}/T \rceil\right)}{T_{ecp}} S_{s_k s_k}(w) \cos(TN_d w), \tag{3.38}$$

where $S_{s_k s_k}(w)$ is the PSD of the transmitted baseband OQAM/FBMC signal on subcarrier k. Therefore, the PSD of the transmitted baseband ECP-OQAM/FBMC signal can be calculated as

$$S_{s^{ecp} s^{ecp}}(w) = \sum_{k=0}^{K-1} S_{s_k^{ecp} s_k^{ecp}}(w)$$

$$= \frac{TN_{d,ecp}}{2T_{ecp}} S_{ss}(w) \left\{1 + \frac{(-1)^{N_d} 2 N_b \left(2\alpha + \lceil 2T_{cp}/T \rceil\right)}{N_{d,ecp}} \cos(TN_d w)\right\},$$

$$\tag{3.39}$$

where $S_{ss}(w)$ is the PSD of the transmitted baseband OQAM/FBMC signal in Eq. (3.24). It can be seen from Eq. (3.39) that after normalization of the coefficient $TN_{d,ecp}/2T_{ecp}$, the ratio of the absolute difference of the PSDs between the transmitted ECP-OQAM/FBMC and OQAM/FBMC signals to the PSD of the transmitted OQAM/FBMC signal is

$$\gamma = \left| \frac{(-1)^{N_d} 2N_b \left(2\alpha + \lceil 2T_{cp}/T \rceil\right)}{N_{d,ecp}} \cos(TN_d w) \right|$$

$$\leq \frac{2N_b \left(2\alpha + \lceil 2T_{cp}/T \rceil\right)}{N_{d,ecp}} \tag{3.40}$$

$$= \frac{2 \left(2\alpha + \lceil 2T_{cp}/T \rceil\right)}{2N_d + 2\alpha + \lceil 2T_{cp}/T \rceil}.$$

The absolute difference of the PSDs (dB) between the transmitted ECP-OQAM/FBMC and OQAM/FBMC signals is then $10\log_{10}(1+\gamma)$, which is 0.3 dB when $\gamma = 1$. Since γ is often less than 1, it indicates that the PSD performance of the transmitted ECP-OQAM/FBMC signal should be nearly as good as that of the original transmitted OQAM/FBMC signal.

3.4 SIMULATION RESULTS

In this section, simulations are conducted to verify the derivations of PSDs and compare the PSD and BER performances of different communication systems. In the following simulations, we set the number of subcarriers as $K = 64$, the modulation method as 4-QAM, the number of blocks as $N_b = 10$, the half block size of input data symbols as $N_d = 20$ and the length of CP as 16. In PSD simulations, the number of realizations is 1000. We use the Welch's method to estimate the PSD, where the window is Blackman window and the overlap is 50%. In BER simulations, perfect synchronization is assumed, channels are known and the minimum mean square error (MMSE) equalizer is used. A random channel model with 15 sample-spaced independent Rayleigh fading paths is adopted in the simulations, which has an exponential power delay profile as $\alpha(n) = e^{-an}$, where $a = 0.2$ and $n = 0, 1, \ldots, 15$. The prototype filter $h(n)$ used in the simulations is given as [7], such as,

$$h(n) = P(0) + 2\sum_{i=1}^{3}(-1)^i P(i) \cos\left(\frac{2\pi i}{4N}n\right), \quad n = 0, 1, \ldots, 4K, \tag{3.41}$$

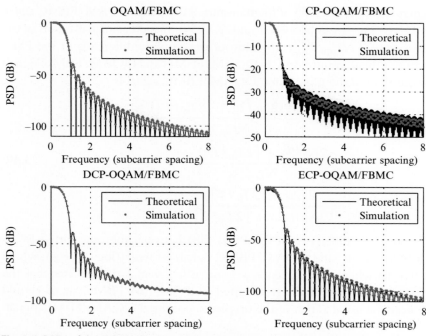

Fig. 3.5 PSDs of the OQAM/FBMC and three CP-based OQAM/FBMC signals.

where $4K + 1$ is the length of the filter with $\alpha = 4$ and $P(0) = 1$, $P(1) = 0.97195983$, $P(2) = 0.70710678$, $P(3) = 0.23514695$.

In Fig. 3.5, the theoretical and simulated PSDs of the transmitted OQAM/FBMC, CP-OQAM/FBMC, DCP-OQAM/FBMC, and ECP-OQAM/FBMC signals are presented. Since the PSD of the transmitted signal is the sum of PSDs on all subcarriers and the PSD on each subcarrier is the frequency shifted version of PSD on subcarrier 0, we only transmit data on subcarrier 0 to verify the theoretical results. The simulated PSDs are obtained by inputting random symbols to the OQAM/FBMC, CP-OQAM/FBMC, DCP-OQAM/FBMC, and ECP-OQAM/FBMC communication systems periodically, then, estimating and averaging the PSDs of the transmitted signal of each communication system. It can be seen from the figures that the theoretical and simulation results match well, which can verify the derivations of PSDs.

In the following simulations, we compare the PSD and BER performances of the CP-based OQAM/FBMC communication systems with those of the windowed CP-OFDM communication system. We transmit data symbols on subcarriers 0 to 31 to compare the mainlobe and sidelobes. In the windowed CP-OFDM communication system, the raised cosine window is used as

$$h_w(t) = \begin{cases} 0.5 + 0.5\cos(\pi + \frac{\pi t}{\beta T}), & 0 \le t \le \beta T, \\ 1, & \beta T \le t \le T, \\ 0.5 + 0.5\cos\left[\frac{\pi(t-T)}{\beta T}\right], & T \le t \le (1+\beta)\,T, \end{cases} \quad (3.42)$$

where T is the symbol duration length and β is the roll-off factor. The details of the windowed CP-OFDM communication system can be found in [8]. Note that when $\beta = 0$, $h_w(t)$ becomes a rectangular pulse and the windowed CP-OFDM communication system becomes the CP-OFDM communication system.

In Fig. 3.6, the BER performances of the CP-based OQAM/FBMC communication systems and windowed CP-OFDM communication systems with different β are presented, where no forward error correction coding is used. The signal-to-noise ratio is defined as E_b/N_0, where

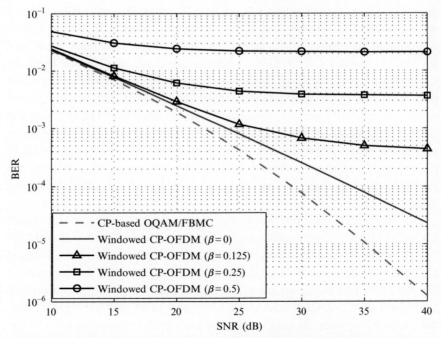

Fig. 3.6 BER performances of the CP-based OQAM/FBMC communication systems and windowed CP-OFDM communication systems with different β.

E_b is the energy per bit excluding the CP and N_0 is the noise spectral density. The three CP-based OQAM/FBMC communication systems have the same BER performances because they have the same system structures and the only differences are the input data symbols or the truncation of the transmitted signal. Since symbols are overlapped in OQAM/FBMC communication systems, CP-based OQAM/FBMC communication systems can achieve more multipath diversity gain than the CP-OFDM communication system when the MMSE equalizer is used. Therefore, CP-based OQAM/FBMC communication systems have better BER performance than the CP-OFDM communication system. Note that a block of data symbols are overlapped in CP-based OQAM/FBMC communication systems and they have to be demodulated and equalized together, while data symbols can be demodulated and equalized independently in the CP-OFDM communication system. The FFT size is then KN_d for CP-based OQAM/FBMC communication systems and K for the CP-OFDM communication system in the equalization process. Therefore, the complexity is $O(KN_d \log(KN_d))$ for CP-based OQAM/FBMC communication systems and $O(KN_d \log K)$ for the CP-OFDM communication system when a block of KN_d complex input data symbols are equalized. All modulation schemes have the same CP length, which is sufficient to avoid the ISI caused by the multipath channel. However, for the windowed CP-OFDM communication system, part of the CP-OFDM symbol is interfered by the window and the length of the interfered part is $\beta(T + T_{cp})$. When β gets larger, the CP-OFDM symbol suffers more interference from the window. It can be seen from Fig. 3.6 that the BER performance of the windowed CP-OFDM communication system gets worse when β gets larger. When $\beta = 0.125, 0.25$, and 0.5, the BER performances of the windowed CP-OFDM communication systems are much worse than those of the CP-based OQAM/FBMC communication systems.

In Fig. 3.7, the PSDs of the transmitted signals in CP-based OQAM/FBMC communication systems and windowed CP-OFDM communication systems with different β are presented. In CP-based OQAM/FBMC communication systems, the sidelobes of the transmitted CP-OQAM/FBMC signal are much higher than those of the transmitted OQAM/FBMC, DCP-OQAM/FBMC, and ECP-OQAM/FBMC signals, which is mainly caused by the truncation in the equivalent form of the CP-OQAM/FBMC communication system. The sidelobes of the transmitted DCP-OQAM/FBMC signal are higher than those of the

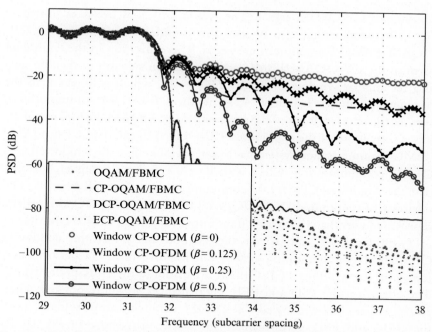

Fig. 3.7 PSDs of the transmitted signals in CP-based OQAM/FBMC communication systems and windowed CP-OFDM communication systems with different β.

transmitted OQAM/FBMC signals due to the CP insertion. However, since the overhead and power of the CP are both small, the sidelobes of the transmitted DCP-OQAM/FBMC signal are still very low, and the PSD is nearly the same as that of the transmitted OQAM/FBMC signal during subcarriers 31–33. Since the transmitted ECP-OQAM/FBMC signal is equivalent to the output signal obtained by inputting cyclically prefixed and suffixed data symbols to an OQAM/FBMC transmitter, the PSD of the transmitted ECP-OQAM/FBMC signal is nearly the same as that of the transmitted OQAM/FBMC signal. As shown in the figure, the curves of the PSDs of the transmitted OQAM/FBMC and ECP-OQAM/FBMC signals are so close that one cannot see obvious difference between them. For the windowed CP-OFDM communication system, the PSD performance gets better when β gets larger. The CP-OFDM communication system has the highest sidelobes. Due to the cyclic structure (or truncation) of the transmitted CP-OQAM/FBMC signal for each block, one might think that its PSD would be similar to the transmitted CP-OFDM signal. The reason why it is better as shown in Fig. 3.7 is that several data symbols are considered together in the CP-OQAM/FBMC communication system. The truncation in the equivalent form of the CP-OQAM/FBMC communication system can then be seen as a rectangular window of length $TN_d + T_{cp}$, which is longer than the rectangular window length $T + T_{cp}$ in the CP-OFDM communication system, thus the sidelobes of the transmitted CP-OQAM/FBMC signal are lower than those of the transmitted CP-OFDM signal. The CP-OQAM/FBMC communication system has lower sidelobes than the windowed CP-OFDM communication system with $\beta = 0.125$, but has higher sidelobes than the windowed CP-OFDM communication system with $\beta = 0.5$. The sidelobes of transmitted DCP-OQAM/FBMC and ECP-OQAM/FBMC signals are lower than those of all transmitted windowed CP-OFDM signals.

From Figs. 3.6 and 3.7, it can be concluded that the windowed CP-OFDM communication system can improve the PSD performance. However, the windowed CP-OFDM is harmful to the BER performance. The effects on the PSD and BER performances are more obvious when β gets larger. For windowed CP-OFDM communication systems, both the BER and PSD performances are worse than those of the DCP-OQAM/FBMC and ECP-OQAM/FBMC communication systems when $\beta = 0, 0.125, 0.25$, and 0.5. For the CP-OQAM/FBMC communication system, it has better BER and PSD performances than the CP-OFDM communication system. When β increases, the PSD performance of the CP-OQAM/FBMC communication system can be outperformed by the windowed CP-OFDM communication system with the cost of BER performance degradation. Thus, considering both the BER and PSD performances, CP-based OQAM/FBMC communication systems perform better than the windowed CP-OFDM communication system. In CP-based OQAM/FBMC communication systems, the ECP-OQAM/FBMC communication system has the best PSD performance. The ratios of the length of original data symbols to the total transmitted signal length are $\frac{T}{T+T_{cp}} = 80.00\%$ for windowed CP-OFDM communication systems, $\frac{TN_d}{TN_d+T_{cp}} = 98.77\%$ for the CP-OQAM/FBMC communication system, $\frac{TN_d}{TN_d+\alpha T - T/2 + T_{cp}} = 84.21\%$ for the DCP-OQAM/FBMC communication system and $\frac{N_b TN_d}{N_b(TN_d+\alpha T+\lceil \frac{2T_{cp}}{T}\rceil \frac{T}{2})+\alpha T - T/2} = 80.48\%$ for the ECP-OQAM/FBMC communication system. From the aspect of data utilization efficiency, CP-based OQAM/FBMC communication systems perform better than windowed CP-OFDM communication systems. However, note that if the windowed CP-OFDM symbols are grouped like the CP-OQAM/FBMC communication system and one CP is added for one whole group of several data blocks as a single OFDM symbol, the data utilization efficiencies of the windowed CP-OFDM and CP-OQAM/FBMC communication systems are the same.

3.5 SUMMARY

In this chapter, we investigated the properties and PSDs of OQAM/FBMC, CP-OQAM/FBMC, DCP-OQAM/FBMC, and ECP-OQAM/FBMC communication systems. We showed that the continuous-time transmitted CP-OQAM/FBMC signal after the CP insertion is equivalent to a truncated signal obtained by inserting CP and CS to the discrete-time input data symbols in the OQAM/FBMC system, which simplifies the calculation of the PSD of the transmitted CP-OQAM/FBMC signal and also implies that the PSD of the transmitted CP-OQAM/FBMC signal may be affected by the truncation. We showed that the PR condition for the CP-OQAM/FBMC communication system is actually the same as that for the OQAM/FBMC communication system when the length of the prototype filter is not more than the length of the transmitted CP-OQAM/FBMC signal of each block excluding the CP. Simulation results show that CP-based OQAM/FBMC communication systems perform better than windowed CP-OFDM communication systems considering both the PSD and BER performances, and the ECP-OQAM/FBMC communication system has the best PSD performance in existing CP-based OQAM/FBMC communication systems with an increase of the data overhead.

REFERENCES

[1] Gao X, Wang W, Xia XG, Au EKS, You X. Cyclic prefixed OQAM-OFDM and its application to single-carrier FDMA. IEEE Trans Commun 2011;59(5):1467–80.
[2] Abdoli MJ, Jia M, Ma J. Weighted circularly convolved filtering in OFDM/OQAM. In: 24th International symposium on personal indoor and mobile radio communications, London, UK; 2013.
[3] Lin H, Siohan P. A new transceiver system for the OFDM/OQAM modulation with cyclic prefix. In: IEEE 19th international symposium on personal, indoor and mobile radio communications (PIMRC); 2008.
[4] Vaidyanathan PP, Kirac A. Theory of cyclic filter banks. In: IEEE international conference on acoustics, speech, and signal processing, Munich, Germany; 1997.

[5] Chen D, Xia XG, Jiang T, Gao X. Properties and power spectral densities of CP based OQAM-OFDM systems. IEEE Trans Signal Process 2015;63(14):3561–75.

[6] Skrzypczak A, Siohan P, Chotkan N, Djoko-Kouam M. OFDM/OQAM: an appropriate modulation scheme for an optimal use of the spectrum. In: 3rd International symposium on communications, control and signal processing, St. Julian's, Malta; 2008.

[7] Mirabbasi S, Martin K. Overlapped complex-modulated transmultiplexer filters with simplified design and superior stopbands. IEEE Trans Circuits Syst II Analog Digit Signal Process 2003;50(8):456–69.

[8] Weiss T, Hillenbrand J, Krohn A, Jondral FK. Mutual interference in OFDM-based spectrum pooling systems. In: 59th vehicular technology conference, Milan, Italy; 2004.

CHAPTER 4

Prototype Filter

Offset quadrature amplitude modulation-based filter bank multicarrier (OQAM/FBMC) techniques utilize filter-bank-based transmultiplexers [1] to channelize the wide signal band. The filter banks, consisting of synthesis and analysis filters, are typically derived from a prototype filter that determines the system performance, such as stopband attenuation, intersymbol interference (ISI) and intercarrier interference (ICI). For the prototype filter design, it is often required that the stopband energy of the filter is minimized, while the nearly perfect reconstruction (NPR) condition is satisfied, that is, the ISI/ICI resulted from the filters is kept lower than a certain threshold. There are also investigations on filters of perfect reconstruction (PR) condition [2, 3] and ISI-free filters in ISI channels [4, 5]. However, the cost of the PR condition and ISI-free property is the increase of stopband attenuation. Excellent frequency selectivity is very important for wireless communication systems, especially for cognitive radio communication systems that rely on the filters for both data transmission and spectrum sensing [6].

The prototype filter optimization methods are categorized into three types in [7]: frequency sampling technique, windowing-based technique, and direct optimization of filter coefficients. Frequency sampling techniques for prototype filter design were proposed in [8, 9]. Different optimization criteria for the frequency sampling technique have been investigated in [7, 10]. Windowing-based techniques for the prototype filter design have been presented in [11]. With the frequency sampling and windowing-based methods, prototype filter coefficients can be given using a closed-form representation that includes few adjustable design parameters. For example, the windowing-based method optimizes the cut-off frequency and the weights of several cosine terms of the filter. On the contrary, direct optimization of prototype filter impulse-response coefficients aims at optimizing all the possible parameters that can affect the performance of the filter, thus has a potential to obtain better performance

OQAM/FBMC for Future Wireless Communications
http://dx.doi.org/10.1016/B978-0-12-813557-0.00004-8

© 2018 Elsevier Ltd.
All rights reserved.

than the frequency sampling and windowing-based methods. However, an evident drawback of this approach is that the number of unknowns (filter coefficients) increases dramatically when the number of subchannels grows high [12]. The general-purposed optimal FIR filter designs can be formulated as convex problems and be efficiently solved with some recently presented algorithms [13, 14]. However, when incorporated with the NPR condition, the problem of direct optimization of filter coefficients is often nonconvex and highly nonlinear, which is very sensitive to initial values and prohibited in practice due to high complexity. Moreover, the global optimality is not guaranteed, because the solution can be easily trapped in a local minimum value [12].

In this chapter, we introduce the three typical filter design methods. We then present the problem formulation of the filter coefficients to both minimize the stopband energy and constrain the ISI/ICI for OQAM/FBMC communication systems. Especially, we take the direct optimization of filter coefficients as an example and present a method to dramatically reduce the number of unknowns of the optimization problem through approximation of the constraints, so that the optimal solution of the approximated optimization problem can be obtained with acceptable computational complexity [15].

The rest of the chapter is organized as follows. In Section 4.1, three filter design methods are briefly introduced. In Section 4.2, we formulate the optimization problem of prototype filter design. In Section 4.3, the number of variables of the optimization problem is reduced and the solution is obtained. In Section 4.4, simulation results are presented. Finally, summary is made in Section 4.5.

4.1 FILTER DESIGN METHODS

The prototype filter is the key of the OQAM/FBMC communication system. As mentioned earlier, the PR condition is determined by the prototype filter design. If the prototype filter is not well designed such that the PR condition is not satisfied, the OQAM/FBMC communication system will suffer the resulted ISI and ICI. Moreover, the prototype filter determines the power spectral density of the OQAM/FBMC signal. Therefore, the main performance of an OQAM/FBMC communication system is determined by the prototype filter.

A prototype filter can be designed to satisfy the PR condition or the NPR condition that causes the ISI and ICI. However, the PR condition is only obtained under an ideal wireless channel, and the ISI/ICI caused by the NPR condition is much smaller than the ISI caused by a practical wireless channel. Moreover, the PR condition may limit the sidelobe performance of the prototype filter. Therefore, the PR condition is not essential for the OQAM/FBMC communication system. It is reasonable to design the prototype filter under the NPR condition and limit the caused ISI/ICI to a low level, which may improve the sidelobe performance of the prototype filter.

Generally, there exist three typical filter design methods, which will be briefly introduced as follows.

- Frequency sampling technique [8]: The key idea is to design the sampling points of the prototype filter in the frequency domain and obtain the prototype filter in the time domain through inverse Fourier transform. The frequency sampling-based filter can be expressed as

$$h(n) = P(0) + 2\sum_{i=1}^{\alpha-1}(-1)^i P(i) \cos\left(\frac{2\pi i}{\alpha K}(n+1)\right), \quad n = 0, 1, \ldots, \alpha K - 2,$$

(4.1)

where $P(i), i = 0, \ldots, \alpha - 1$ are parameters to be optimized. It can be seen from Eq. (4.1) that the number of variables is small and not related to the length of the prototype filter.

- Windowing technique [11]: The key idea is to multiply a window function to an existing low-pass filter to obtain the prototype filter. The windowing-based filter can be expressed as

$$h(n) = w(n)h_c(n),$$

(4.2)

where $h_c(n)$ and $w(n)$ are

$$h_c(n) = \frac{\sin\left[w_c(n - (L_p - 1)/2)\right]}{\pi(n - (L_p - 1)/2)}, \quad n = 0, 1, \ldots, L_p - 1,$$

(4.3)

$$w(n) = \sum_{i=0}^{3}(-1)^i A_i \cos\left(\frac{2\pi i n}{L_p - 1}\right),$$

(4.4)

respectively, where w_c is the cutoff frequency, and $A(i), i = 0, 1, 2, 3$ are the weights of four cosine items. It can be seen from Eq. (4.4) that the number of variables is also small for the windowing-based filter.

- Direct optimization of filter coefficients: The direct but complicated filter design technique. In this technique, the prototype filter has no explicit expression. Since there are typically hundreds of filter coefficients and the NPR condition is nonconvex, it is hard to obtain the

Globally optimized prototype filter. Nevertheless, the direct optimization technique has the potential to obtain the best sidelobe performance
of the prototype filter if the filter optimization problem can be efficiently
solved.

For the above three filter design methods, there are typically three
optimization criteria, which are as follows.

- Minimize the stopband energy: Since the first sidelobe dominates the
  overall stopband energy, this criterion is aimed to minimize the overall
  energy leaked to other frequency bands.
- Minimize the maximum stopband magnitude: The aim is to suppress the
  sidelobes near the stopband edge. However, the stopband energy will be
  increased.
- Peak constrained least square: It is the tradeoff between the previous
  two criteria. The aim is to minimize the total energy during the
  predetermined frequency interval.

In filter optimization problems with the above three criteria, the
constraint is often to limit the ISI/ICI power caused by the NPR condition.

4.2 PROBLEM FORMULATION

In this section, we first analyze the ISI/ICI caused by the NPR condition,
then, formulate the optimization problem of the prototype filter design.

4.2.1 ISI/ICI Caused by the NPR Condition

Denote the ISI/ICI interference to $a_k(m)$ and $b_k(m)$ by $I^a_{k,m}$ and $I^b_{k,m}$,
respectively. According to Eqs. (2.4), (2.5), the expected power of $I^a_{k,m}$ and
$I^b_{k,m}$ are

$$
\begin{aligned}
\text{Power}(I^a_{k,m}) &= \mathrm{E}\left[\left(\hat{a}_k(m) - a_k(m)\right)^2\right] \\
&= \mathrm{E}\left[\left(\sum_{m'=-\infty}^{\infty} \sum_{k'=0}^{K-1} \int_{-\infty}^{\infty} h\left(mT - t\right)\left\{a_{k'}(m')h(t - m'T)\cos\left[(k' - k)\varphi_t\right]\right.\right.\right. \\
&\qquad\qquad \left.\left.\left. -b_{k'}(m')h\left(t - m'T - T/2\right)\sin\left[(k' - k)\varphi_t\right]\right\} dt - a_k(m)\right)^2\right] \\
&= \mathrm{E}\left[\left(\sum_{m'=-\infty}^{\infty} \sum_{k'=0}^{K-1} I^a_{k,m,k',m'} - a_k(m)\right)^2\right],
\end{aligned}
\tag{4.5}
$$

$$\text{Power}(I^b_{k,m}) = \text{E}\left[\left(\hat{b}_k(m) - b_k(m)\right)^2\right]$$

$$= \text{E}\left[\left(\sum_{m'=-\infty}^{\infty}\sum_{k'=0}^{K-1} I^b_{k,m,k',m'} - b_k(m)\right)^2\right], \quad (4.6)$$

respectively, where $I^a_{k,m,k',m'}$ represents the contribution of $x_{k'}(m')$ to $\hat{a}_k(m)$ and $I^b_{k,m,k',m'}$ represents the contribution of $x_{k'}(m')$ to $\hat{b}_k(m)$. According to Eq. (2.4), the contribution of $x_{k'}(m')$ to $\hat{a}_k(m)$ is

$$I^a_{k,m,k',m'} = a_{k'}(m')C'_{k,m,k',m'} - b_{k'}(m')C''_{k,m,k',m'}, \quad (4.7)$$

where

$$C'_{k,m,k',m'} = \int_{-\infty}^{\infty} h(mT-t)h(t-m'T)\cos\left[(k'-k)\varphi_t\right] dt, \quad (4.8)$$

$$C''_{k,m,k',m'} = \int_{-\infty}^{\infty} h(mT-t)h\left(t-m'T-\frac{T}{2}\right)\sin\left[(k'-k)\varphi_t\right] dt. \quad (4.9)$$

To design the filter in the discrete time domain, we replace $h(t)$ with its discrete time version $h(l), l = 0, 1, \ldots, L_p - 1$ for Eqs. (4.8), (4.9), where $h(n)$ corresponds to the filter impulse response at time lT/K and L_p represents the length of the discrete time filter. The discrete time expressions of Eqs. (4.8), (4.9) are then

$$C'_{k,m,k',m'} = \sum_{n=0}^{L_p-1}\left\{h(mK-n)h(n-m'K)\cos\left[(k'-k)(2\pi n/K + \pi/2)\right]\right\}, \quad (4.10)$$

$$C''_{k,m,k',m'} = \sum_{n=0}^{L_p-1}\left\{h(mK-n)h(n-m'K-K/2)\right.$$
$$\left.\sin\left[(k'-k)(2\pi n/K + \pi/2)\right]\right\}, \quad (4.11)$$

respectively.

The distributions of $a_{k'}(m')$ and $b_{k'}(m')$ have unit power and are symmetric to the origin due to randomness of the information bits, which can be expressed as

$$\text{E}[(a_{k'}(m'))^2] = \text{E}[(b_{k'}(m'))^2] = 1, \quad (4.12)$$

$$\text{E}[a_{k'}(m')] = \text{E}[b_{k'}(m')] = 0. \quad (4.13)$$

Moreover, the distributions of $a_{k'}(m')$ and $b_{k'}(m')$ are independent. Then, from Eqs. (4.7), (4.12), (4.13), the expectation of the power of $I^a_{k,m,k',m'}$ is given as

$$
\begin{aligned}
\mathrm{E}\left[\left(I^a_{k,m,k',m'}\right)^2\right] &= \left(C'_{k,m,k',m'}\right)^2 \mathrm{E}\left[(a_{k'}(m'))^2\right] + \left(C''_{k,m,k',m'}\right)^2 \mathrm{E}\left[(b_{k'}(m'))^2\right] \\
&\quad - 2C'_{k,m,k',m'} C''_{k,m,k',m'} \mathrm{E}[a_{k'}(m')]\mathrm{E}[b_{k'}(m')] \\
&= \left(C'_{k,m,k',m'}\right)^2 + \left(C''_{k,m,k',m'}\right)^2.
\end{aligned}
\tag{4.14}
$$

By evaluating Eqs. (4.10), (4.11) with $(k',m') = (k,m)$, we have $C'_{k,m,k,m} = \sum_{n=0}^{L_p-1}(h(n))^2$ and $C''_{k,m,k,m} = 0$. Therefore, the contribution of $x_k(m)$ to $\hat{a}_k(m)$ can be expressed as

$$
\begin{aligned}
I^a_{k,m,k,m} &= a_k(m) C'_{k,m,k,m} - b_k(m) C''_{k,m,k,m} \\
&= a_k(m) \sum_{n=0}^{L_p-1}(h(n))^2.
\end{aligned}
\tag{4.15}
$$

According to Eq. (4.15), in order to recover $a_k(m)$ at the receiver, it should be satisfied that

$$
\sum_{n=0}^{L_p-1}(h(n))^2 = 1,
\tag{4.16}
$$

$$
I^a_{k,m,k,m} = a_k(m).
\tag{4.17}
$$

Note that the distributions of $a_{k'}(m')$ and $b_{k'}(m')$ are independent for different (k',m'). Thus, the expected power of the ISI/ICI to $a_k(m)$ is

$$
\begin{aligned}
\mathrm{Power}(I^a_{k,m}) &= \mathrm{E}\left[\left(\sum_{m'=-\infty}^{\infty}\sum_{k'=0}^{K-1} I^a_{k,m,k',m'} - a_k(m)\right)^2\right] \\
&= \sum_{m'=-\infty}^{\infty}\left\{\sum_{k'=0,(k',m')\neq(k,m)}^{K-1} \mathrm{E}\left[(I^a_{k,m,k',m'})^2\right]\right\} \\
&\quad + \mathrm{E}\left[(I^a_{k,m,k,m} - a_k(m))^2\right] \\
&= \sum_{m'=-\infty}^{\infty}\left\{\sum_{k'=0,(k',m')\neq(k,m)}^{K-1} \mathrm{E}\left[(I^a_{k,m,k',m'})^2\right]\right\}.
\end{aligned}
\tag{4.18}
$$

Due to the symmetry property, the expected power of the ISI/ICI to $b_k(m)$, Power($I^b_{k,m}$), can be similarly calculated and is equal to Power($I^a_{k,m}$). The major steps of the proof are given as follows.

First, we transform Eq. (2.5) as

$$
\hat{b}_k(m) = \sum_{k'=0}^{K-1} \int_{-\infty}^{\infty} h\left(mT - t + T/2\right)
$$

$$
\times \left\{ \sum_{m'=-\infty}^{\infty} b_{k'}(m')h\left(t - m'T - T/2\right) \cos\left[(k' - k)\varphi_t\right] \right.
$$

$$
\left. + \sum_{m'=-\infty}^{\infty} a_{k'}(m')h(t - m'T) \sin\left[(k' - k)\varphi_t\right] \right\} dt. \tag{4.19}
$$

Noticing that the summation over m' in Eq. (4.19) is from $-\infty$ to ∞, we can replace m' with $m' + 1$ in the second summation in braces. Eq. (4.19) can then be rewritten as

$$
\hat{b}_k(m) = \sum_{k'=0}^{K-1} \int_{-\infty}^{\infty} h\left(mT - t + T/2\right)
$$

$$
\times \left\{ \sum_{m'=-\infty}^{\infty} b_{k'}(m')h\left(t - m'T - T/2\right) \cos\left[(k' - k)\varphi_t\right] \right.
$$

$$
\left. + \sum_{m'=-\infty}^{\infty} a_{k'}(m' + 1)h(t - m'T - T) \sin\left[(k' - k)\varphi_t\right] \right\} dt. \tag{4.20}
$$

The integral over t in Eq. (4.20) in is also from $-\infty$ to ∞. We can then replace t with $t + T/2$ in Eq. (4.20) and obtain

$$
\hat{b}_k(m) = \sum_{m'=-\infty}^{\infty} \sum_{k'=0}^{K-1} \int_{-\infty}^{\infty} h\left(mT - t\right)
$$

$$
\times \left\{ b_{k'}(m')h\left(t - m'T\right) \cos\left[(k' - k)\varphi_{t+T/2}\right] \right.
$$

$$
\left. + a_{k'}(m' + 1)h(t - m'T - T/2) \sin\left[(k' - k)\varphi_{t+T/2}\right] \right\} dt. \tag{4.21}
$$

Since $\varphi_{t+T/2} = \frac{2\pi(t+T/2)}{T} + \frac{\pi}{2} = \varphi_t + \pi$, Eq. (4.21) can be rewritten as

$$\hat{b}_k(m) = \sum_{m'=-\infty}^{\infty} \sum_{k'=0}^{K-1} \int_{-\infty}^{\infty} h(mT - t)$$

$$\times \left\{ (-1)^{k'-k} b_{k'}(m')h\left(t - m'T\right)\cos\left[(k' - k)\varphi_t\right] \right.$$

$$\left. + (-1)^{k'-k} a_{k'}(m'+1)h(t - m'T - T/2)\sin\left[(k' - k)\varphi_t\right] \right\} dt.$$

$$(4.22)$$

It can be observed that Eqs. (4.22), (2.4) are symmetric regardless of the item $(-1)^{k'-k}$. Exploit this symmetry and follow the steps of Eqs. (4.6)–(4.18), it is easy to derive that Power$(I_{k,m}^b) = $ Power$(I_{k,m}^a)$.

Since all data symbols are independent and identically distributed in OQAM/FBMC communication systems, it can be seen that Power$(I_{k,m}^a)$ and Power$(I_{k,m}^b)$ are independent of k and m. Therefore, the power level of the total ISI/ICI of an OQAM/FBMC communication system can be measured by Power$(I_{k,m}^a)$ and Power$(I_{k,m}^b)$ with any choice of k and m.

4.2.2 Direct Optimization of Filter Coefficients

Typically, $h(n)$ is an even and real NPR filter with the length L_p. Then, $h(n)$ should satisfy

$$h(n) = h(L_p - 1 - n), \quad n = 0, 1, \ldots, L_p - 1. \qquad (4.23)$$

The Fourier transform of the designed filter $h(n)$ is

$$H(e^{jw}) = \sum_{n=0}^{L_p-1} h(n)e^{-jwn}. \qquad (4.24)$$

The magnitude response of the filter $h(n)$ is

$$|H(e^{jw})| = \left| \sum_{n=0}^{L_p-1} h(n)e^{-jwn} \right|. \qquad (4.25)$$

The optimization objective is to minimize the stopband energy of the prototype filter, where the stopband region is denoted by $[w_0, \pi]$. With the constraint of Eq. (4.16) and constraint of the ISI/ICI power, the filter design problem can be written as

$$\mathbf{P1:} \quad \min_{h(0),h(1),\ldots,h(L_p-1)} \int_{w_0}^{\pi} |H(e^{jw})|^2 \, dw, \tag{4.26a}$$

$$\text{subject to} \quad h(n) = h(L_p - 1 - n), \quad n = 0, 1, \ldots, L_p - 1, \tag{4.26b}$$

$$\text{Power}(I_{k,m}^a) \leq \text{TH}, \tag{4.26c}$$

$$\text{Power}(I_{k,m}^b) \leq \text{TH}, \tag{4.26d}$$

$$\sum_{n=0}^{L_p-1} (h(n))^2 = 1, \tag{4.26e}$$

where a low threshold TH guarantees that the error between the input at the transmitter and the output at the receiver is small enough so that the designed filter is an NPR filter. Since $\text{Power}(I_{k,m}^a) = \text{Power}(I_{k,m}^b)$, the constraint (4.26d) can be removed directly. Moreover, the constraint (4.26e) can be relaxed as

$$\sum_{n=0}^{L_p-1} (h(n))^2 \geq 1. \tag{4.27}$$

The proof of the relaxation (4.27) is as follows. Assume $h'(n)$ is the optimal solution of **P1** with the relaxed constraint and satisfies $\sum_{n=0}^{L_p-1} (h'(n))^2 = a^2 > 1$, where a is a constant. Then, it is obvious that $h^\star(n) = \frac{1}{a}h'(n)$ also satisfies constraints (4.26b)–(4.26d), (4.27). However, the objective value (4.26a) for $h^\star(n)$ is smaller, which contradicts the fact that $h'(n)$ is the optimal solution. Therefore, the optimal solution must satisfy $\sum_{n=0}^{L_p-1} (h'(n))^2 = 1$.

Moreover, we could control the stopband attenuation within specified frequency ranges by adding weights to the objective function (4.26a). Then, the optimization problem **P1** is modified as

$$\mathbf{P2:} \quad \min_{h(0),h(1),\ldots,h(L_p-1)} \int_{w_0}^{\pi} W(w)|H(e^{jw})|^2 \, dw, \tag{4.28a}$$

$$\text{subject to} \quad h(n) = h(L_p - 1 - n), \quad n = 0, 1, \ldots, L_p - 1, \tag{4.28b}$$

$$\text{Power}(I_{k,m}^a) \leq \text{TH}, \tag{4.28c}$$

$$\sum_{n=0}^{L_p-1} (h(n))^2 \geq 1, \tag{4.28d}$$

where $W(w) > 0$ is the weight of frequency w.

According to Eq. (4.28b), only half of the variables are independent in **P2**. Thus, let

$$\mathbf{x} = [x_1, x_2, \ldots, x_L]^T$$

$$= \begin{cases} \left[h(0), h(1), \ldots, h\left(\frac{L_p}{2} - 1\right)\right]^T, & \text{if } L_p \text{ is even,} \\ \left[h(0), h(1), \ldots, h\left(\frac{L_p-1}{2}\right)\right]^T, & \text{if } L_p \text{ is odd.} \end{cases} \quad (4.29)$$

The optimization problem **P2** can then be rewritten as

$$\mathbf{P3:} \ \min_{\mathbf{x}} \ f_0(\mathbf{x}) = \int_{w_0}^{\pi} W(w)|H(e^{jw})|^2 dw, \quad (4.30a)$$

$$\text{subject to} \quad f_1(\mathbf{x}) = \text{Power}(I_{k,m}^a) - \text{TH} \le 0, \quad (4.30b)$$

$$f_2(\mathbf{x}) = -\left(\sum_{n=0}^{L_p-1} (h(n))^2 - 1\right) \le 0. \quad (4.30c)$$

Unfortunately, the problem **P3** is nonconvex due to the nonconvexity of (4.30b), (4.30c). Nevertheless, considering that all the functions in **P3** are twice-differentiable, it is available to apply the branch and bound algorithm to this kind of problem.

4.2.3 Branch and Bound Algorithm

The branch and bound algorithm offers mathematical guarantees for the convergence to a point arbitrarily close to the global minimum for the large class of twice-differentiable nonlinear programming problems. The key idea is to construct a converging sequence of upper and lower bounds on the global minimum through the convex relaxation of the original problem [16–18].

A typical way to obtain the tight convex relaxation is to apply the α-based branch and bound (αBB) algorithm. Each iteration of the αBB algorithm consists of a branching step and a bounding step. Take **P3** as an example. In the branching steps, the set of constraints in **P3** is split into two smaller subsets. In the bounding steps, the upper and lower bounds for the minimum value of $f_0(\mathbf{x})$ within the given subsets are computed. The lower bound is obtained by constructing valid convex underestimators for the functions in the problem **P3**, using the convex underestimators to relax the problem **P3** to a convex optimization and solving the convex optimization. The upper bound is obtained by searching a feasible solution of the original problem locally within the given subset. A global variable is used to record the minimum upper bound obtained among all subsets examined so far. If the lower bound for a subset is greater than the minimum upper bound, the subset can be safely discarded. The iteration stops when the lower bounds of the remaining subsets match the minimum upper bound and the feasible solution corresponding to the minimum upper bound is actually the global optimal solution of the original problem.

Let $[\mathbf{x}^L, \mathbf{x}^U]$ denote a box constraint of $\{\mathbf{x} | x_j^L \leq x_j \leq x_j^U, \ j = 1, 2, \ldots, L\}$, where $\mathbf{x}^L = (x_1^L, x_2^L, \ldots, x_L^L)^T$ and $\mathbf{x}^U = (x_1^U, x_2^U, \ldots, x_L^U)^T$. We choose a box that is sufficiently large to contain the constraints (4.30b), (4.30c) as the initial set. In the branching step, we split a subset equally into two subsets along x_i that has the longest range $|x_j^L - x_j^U|$. In the bounding steps, $f_i(\mathbf{x})$ is underestimated over the domain $[\mathbf{x}^L, \mathbf{x}^U]$ by the function $L_i(\mathbf{x})$ as

$$L_i(\mathbf{x}) = f_i(\mathbf{x}) + \sum_{j=1}^{L} \alpha_i \left(x_j^L - x_j \right) \left(x_j^U - x_j \right), \quad i = 0, 1, 2. \quad (4.31)$$

Maranas and Floudas showed that the underestimation in Eq. (4.31) is tight because the maximum separation distance between $L_i(\mathbf{x})$ and $f_i(\mathbf{x})$ is bounded and converges to zero as $|\mathbf{x}^U - \mathbf{x}^L|$ approaches zero. In order to make $L_i(\mathbf{x})$ convex, α_i should be generated as

$$\alpha_i \geq \max \left\{ 0, -\frac{1}{2} \min_{\mathbf{x} \in [\mathbf{x}^L, \mathbf{x}^U]} \lambda_i(\mathbf{x}) \right\}, \quad (4.32)$$

where $\lambda_i(\mathbf{x})$ represents the eigenvalues of $H_{f_i}(\mathbf{x})$, the Hessian matrix of the function $f_i(\mathbf{x})$. The detailed description of the branch and bound algorithm and the generation of α_i can be found in [17].

Though the optimization problem **P3** can optimally solved by the branch and bound algorithm in theory, the convergence time is too long to be acceptable in practice. Since the convergence time is largely determined by the number of variables (i.e., the filter coefficients) in the optimization, the main purpose of the next section is to reduce the number of variables.

4.3 PROBLEM SOLUTION

4.3.1 Variable Transformation

The objective function (4.30a) is a convex quadratic function, which can be written in vector form as $f_0(\mathbf{x}) = \mathbf{x}^T \mathbf{C} \mathbf{x}$, where $\mathbf{C}$ is a real symmetric positive-definite matrix. Denoting the ith row and jth column element of $\mathbf{C}$ by c_{ij}, then $f_0(\mathbf{x})$ can be expressed as

$$
\begin{aligned}
f_0(\mathbf{x}) = \mathbf{x}^T \mathbf{C} \mathbf{x} &= \sum_{i=1}^{L} \sum_{j=1}^{L} c_{ij} x_i x_j \\
&= \int_{w_0}^{\pi} W(w) |H(e^{jw})|^2 \, dw \\
&= \int_{w_0}^{\pi} W(w) \left| \sum_{n=0}^{L_p-1} h(n) e^{-jwn} \right|^2 dw \\
&= \int_{w_0}^{\pi} W(w) \left\{ \left(\sum_{n=0}^{L_p-1} h(n) \cos(wn) \right)^2 + \left(\sum_{n=0}^{L_p-1} h(n) \sin(wn) \right)^2 \right\} dw.
\end{aligned}
$$

$$(4.33)$$

By setting x_i equal to 1 and all other variables equal to 0 in Eq. (4.33), c_{ii} is calculated as

$$
c_{ii} = f_0(\mathbf{x} = [0, \ldots, x_i = 1, 0, \ldots, 0]^T), \quad i = 1, 2, \ldots, L. \qquad (4.34)
$$

Similarly, c_{ij} $(i \neq j)$ can be obtained by setting x_i, x_j equal to 1 and all other variables equal to 0 in Eq. (4.33), which is expressed as

$$
\begin{aligned}
c_{ij} = 1/2 \times \big\{ f_0(\mathbf{x} = [0, \ldots, x_i = 1, 0, \ldots, x_j = 1, 0, \ldots, 0]^T) \\
- c_{ii} - c_{jj} \big\}, \quad i, j = 1, 2, \ldots, L, \quad i \neq j.
\end{aligned}
$$

$$(4.35)$$

Utilizing Eqs. (4.29), (4.33)–(4.35), the value of c_{ij} can be easily calculated.

Since $\mathbf{C}$ is a real symmetric matrix, we can find L positive eigenvalues $\lambda_1, \lambda_2, \ldots, \lambda_L$ in ascending order $(0 < \lambda_1 \leq \lambda_2 \leq \cdots \leq \lambda_L)$ and the corresponding orthonormal eigenvectors $\mathbf{v}_1, \mathbf{v}_2, \ldots, \mathbf{v}_L$ (column vectors) for $\mathbf{C}$. We then apply an orthonormal transformation to $\mathbf{x}$ as

$$
\mathbf{x} = \mathbf{V} \mathbf{y}, \qquad (4.36)
$$

where $\mathbf{V} = [\mathbf{v}_1, \mathbf{v}_2, \ldots, \mathbf{v}_L]$ is the transformation matrix and $\mathbf{y} = [y_1, y_2, \ldots, y_L]^T$ denotes the transformed variables. With the linear transformation in Eq. (4.36), $f_0(\mathbf{x})$ becomes

$$f_0(\mathbf{x}) = \mathbf{x}^T \mathbf{C} \mathbf{x} = \mathbf{y}^T \mathrm{Diag}(\lambda_1, \lambda_2, \ldots, \lambda_L) \mathbf{y}$$
$$= \lambda_1 y_1^2 + \lambda_2 y_2^2 + \cdots + \lambda_L y_L^2, \qquad (4.37)$$

where $\mathrm{Diag}(\cdot)$ denotes a diagonal matrix. Let

$$g_0(\mathbf{y}) = \lambda_1 y_1^2 + \lambda_2 y_2^2 + \cdots + \lambda_L y_L^2 = f_0(\mathbf{x}). \qquad (4.38)$$

We also perform the linear transformation on the constraints (4.30b), (4.30c) as

$$g_i(\mathbf{y}) = f_i(\mathbf{x}), \quad i = 1, 2. \qquad (4.39)$$

The optimization problem **P3** can then be rewritten as

$$\mathbf{P4}: \min_{\mathbf{y}} g_0(\mathbf{y}),$$
$$\text{subject to} \quad g_i(\mathbf{y}) \le 0, \quad i = 1, 2. \qquad (4.40)$$

4.3.2 Determination of Search Region With a Feasible Solution

In this section, we aim to determine the search region for each variable with the help of a known feasible solution of problem **P4**.

The filter obtained using the frequency sampling technique in [8] is a feasible solution that satisfies all constraints of Eq. (4.26), as long as TH is not set to be too small. The impulse response coefficients of the filter are given as

$$h(n) = P(0) + 2 \sum_{i=1}^{\alpha-1} (-1)^i P(i) \cos\left(\frac{2\pi i}{\alpha K}(l+1)\right), \quad n = 0, 1, \ldots, \alpha K - 2,$$

$$(4.41)$$

where $\alpha K - 1$ is the length of the filter and $P(i)$ are coefficients related to α. For $\alpha = 3$, $P(0) = 1$, $P(1) = 0.91143783$, $P(2) = 0.41143783$; for $\alpha = 4$, $P(0) = 1$, $P(1) = 0.97195983$, $P(2) = 0.70710678$, $P(3) = 0.23514695$.

Let $\mathbf{y}_0$ denote the vector of transformed variables that represents the filter obtained using the frequency sampling technique. Since $\mathbf{y}_0$ is a feasible solution of the problem **P4**, the optimal solution $\mathbf{y}^*$ must satisfy

$$\lambda_1 (y_1^*)^2 + \lambda_2 (y_2^*)^2 + \cdots + \lambda_L (y_L^*)^2 = g_0(\mathbf{y}^*) \le g_0(\mathbf{y}_0). \qquad (4.42)$$

This means that the search region of the optimal solution can be greatly reduced with the knowledge of the feasible solution $\mathbf{y}_0$, that is, the optimization problem **P4** becomes

$$\mathbf{P5}: \min_{\mathbf{y}} g_0(\mathbf{y}),$$

$$\text{subject to} \quad g_i(\mathbf{y}) \leq 0, \quad i = 1, 2, \qquad (4.43)$$

$$\lambda_1 y_1^2 + \lambda_2 y_2^2 + \cdots + \lambda_L y_L^2 \leq g_0(\mathbf{y}_0).$$

The newly added constraint in the optimization problem **P5** infers that

$$\lambda_j (y_j^\star)^2 \leq \lambda_1 (y_1^\star)^2 + \lambda_2 (y_2^\star)^2 + \cdots + \lambda_L (y_L^\star)^2 \leq g_0(\mathbf{y}_0)$$

$$\Longrightarrow \sqrt{\lambda_i} |y_j^\star| \leq \sqrt{g_0(\mathbf{y}_0)}$$

$$\Longrightarrow -\sqrt{\frac{g_0(\mathbf{y}_0)}{\lambda_j}} \leq y_j^\star \leq \sqrt{\frac{g_0(\mathbf{y}_0)}{\lambda_i}}, \quad j = 1, 2, \ldots, L. \qquad (4.44)$$

Therefore, we can use the box constraint (4.44) as the initial set for the branch and bound algorithm.

4.3.3 Reduction of the Variable Number

According to (4.44), it can be seen that the variable y_j with larger λ_j is constrained to be smaller. For λ_j large enough, the variable y_j is so small that it may be neglected for the constraints $g_i(\mathbf{y})$, $i = 1, 2$. Thus, we conjecture that it is a good approximation for all constraints $g_i(\mathbf{y})$ that the variables corresponding to the smaller eigenvalues are preserved while the variables corresponding to the larger eigenvalues are set as zero.

For each variable vector $\mathbf{y} = [y_1, y_2, \ldots, y_L]$, we form a new vector $\mathbf{y}'$ of length L, where $\mathbf{y}' = [y_1, y_2, \ldots, y_{L'}, 0, \ldots, 0]$. If L' is large enough, the constraints can be relaxed as

$$g_i(\mathbf{y}) \approx g_i(\mathbf{y}'), \quad i = 1, 2. \qquad (4.45)$$

The approximation (4.45) and the determination of L' will be discussed in the following section. According to (4.45), **P5** can be approximated as

$$\mathbf{P6}: \min_{\mathbf{y}} g_0(\mathbf{y}) = \left\{ \sum_{j=1}^{L'} \lambda_j y_j^2 + \sum_{j=L'+1}^{L} \lambda_j y_j^2 \right\}, \qquad (4.46a)$$

$$\text{subject to} \quad g_i(\mathbf{y'}) \leq 0, \quad i = 1, 2, \tag{4.46b}$$

$$\sum_{j=1}^{L} \lambda_j \gamma_j^2 \leq g_0(\mathbf{y_0}). \tag{4.46c}$$

Next, we consider the optimization problem after relaxing (4.46c) in **P6** as

$$\textbf{P7:} \; \min_{\mathbf{y}} g_0(\mathbf{y}) = \left\{ \sum_{j=1}^{L'} \lambda_j \gamma_j^2 + \sum_{j=L'+1}^{L} \lambda_j \gamma_j^2 \right\}, \tag{4.47a}$$

$$\text{subject to} \quad g_i(\mathbf{y'}) \leq 0, \quad i = 1, 2, \tag{4.47b}$$

$$\sum_{j=1}^{L'} \lambda_j \gamma_j^2 \leq g_0(\mathbf{y_0}). \tag{4.47c}$$

Let $\mathbf{y}^{\dagger}$ denote the optimal solution of the problem **P7**. Since $\gamma_{L'+1}, \ldots, \gamma_L$ are not presented in the constraints and $\sum_{j=L'+1}^{L} \lambda_j \gamma_j^2 \geq 0$, the optimal solution of **P7** must satisfy that $\gamma_{L'+1}^{\dagger} = \cdots = \gamma_L^{\dagger} = 0$. Therefore, $\mathbf{y}^{\dagger}$ satisfies (4.46c) and it is a feasible solution of the problem **P6**, which means that $g_0(\mathbf{y}^{\dagger}) \geq g_0(\mathbf{y}^{\star})$, where $\mathbf{y}^{\star}$ is the optimal solution of **P6**. Meanwhile, (4.47c) is a relaxation of (4.46c), which means that $g_0(\mathbf{y}^{\dagger}) \leq g_0(\mathbf{y}^{\star})$. Therefore, it is clear that $g_0(\mathbf{y}^{\dagger}) = g_0(\mathbf{y}^{\star})$, which means that optimal solution of **P7** is also the optimal solution of **P6**, and $\gamma_{L'+1}^{\star} = \cdots = \gamma_L^{\star} = 0$.

After removing the zero variables $\gamma_{L'+1}, \ldots, \gamma_L$ in **P7**, **P7** can be rewritten as

$$\textbf{P8:} \; \min_{\gamma_1, \ldots, \gamma_{L'}} \sum_{j=1}^{L'} \lambda_j \gamma_j^2,$$

$$\text{subject to} \quad g_i(\gamma_1, \ldots, \gamma_{L'}, 0, \ldots, 0) \leq 0, \quad i = 1, 2, \tag{4.48}$$

$$\sum_{j=1}^{L'} \lambda_j \gamma_j^2 \leq g_0(\mathbf{y_0}).$$

The problem **P8** can be efficiently solved by the branch and bound algorithm if L' is very small.

4.3.4 Verification of the Approximation of Constraints

To verify that the approximation (4.45) is good for a certain L', we can increase L' gradually and solve **P8** for each L'. If the resulted objective value stops decreasing (or the decrease is negligible) for a certain L' and above, the approximation of preserving this certain number of variables is good enough. The detail will be discussed in the section of simulation results.

As another way to verify that the approximation (4.45) is reasonable for a certain L', we can compute the maximum difference $d_i(L')$ between $g_i(\mathbf{y})$ and $g_i(\mathbf{y}')$ over the search region of **P5**, for $i = 1, 2$, respectively. This leads to the optimization problem as

$$d_i(L') = \max_{\mathbf{y}} |g_i(\mathbf{y}) - g_i(\mathbf{y}')|, \quad i = 1, 2$$

$$\text{subject to} \quad g_1(\mathbf{y}) \leq 0, \quad g_2(\mathbf{y}) \leq 0,$$

$$\sum_{j=1}^{L} \lambda_j \gamma_j^2 \leq g_0(\mathbf{y}_0). \tag{4.49}$$

Since $d_i(L')$ is the maximum difference between constraints $g_i(\mathbf{y})$ and $g_i(\mathbf{y}')$, and $g_i(\mathbf{y}) \leq 0$ is meant to satisfy the NPR condition, approximating $g_i(\mathbf{y})$ with $g_i(\mathbf{y}')$ is reasonable as long as $d_1(L') \ll \text{TH}$ and $d_2(L') \ll 1$.

In order to make the optimization problem (4.49) smooth, we can transform it into two optimization problems as

$$d_i^+(L') = \max_{\mathbf{y}} \left(g_i(\mathbf{y}) - g_i(\mathbf{y}')\right), \quad i = 1, 2,$$

$$\text{subject to} \quad g_1(\mathbf{y}) \leq 0, \quad g_2(\mathbf{y}) \leq 0,$$

$$\sum_{j=1}^{L} \lambda_j \gamma_j^2 \leq g_0(\mathbf{y}_0). \tag{4.50}$$

$$d_i^-(L') = \max_{\mathbf{y}} \left(g_i(\mathbf{y}') - g_i(\mathbf{y})\right), \quad i = 1, 2,$$

$$\text{subject to} \quad g_1(\mathbf{y}) \leq 0, \quad g_2(\mathbf{y}) \leq 0,$$

$$\sum_{j=1}^{L} \lambda_j \gamma_j^2 \leq g_0(\mathbf{y}_0). \tag{4.51}$$

By solving the two sub–problems (4.50), (4.51), we have $d_i(L') = \max\left(d_i^+(L'), d_i^-(L')\right)$.

Note that solving the problem (4.49) is as difficult as solving **P5**. However, suboptimal solutions of (4.49), which are available by applying heuristic algorithms, reflect the magnitude order of $d_i(L')$ and can be used as approximations of $d_i(L')$. One suboptimal algorithm we tried is sequential quadratic programming (SQP) [19]. Since the solution obtained by the SQP algorithm is highly dependent on the initial variables, we apply the SQP algorithm to the optimization problems (4.50), (4.51) multiple times (1000 times in simulations) with random initial variables within the search region, and use the maximum result to approximate $d_i(L')$.

4.4 SIMULATION RESULTS

In this section, simulations are conducted to determine the appropriate number of preserved variables, and compare the sidelobe performances. We set the ISI/ICI threshold TH$= 10^{-3}$ and 10^{-4}, the number of subcarriers $K = 64$ and 256, and the filter length $L_p = 3K - 1$ and $4K - 1$ for the optimization problem. The weight $W(w)$ is set to be 1 for the computations of Figs. 4.2, 4.3, and 4.6, where the optimization objective is to minimize the stopband energy within $[w_0, \pi] = [2\pi/N, \pi]$. $W(w)$ is varied for Figs. 4.4 and 4.5, where the optimization objective is to control the stopband attenuation within specified frequency ranges.

By solving the problem (4.49) with the SQP algorithm, within a range of L', we obtain the corresponding $d_i(L')$ for $i = 1, 2$, respectively. For the SQP algorithm, we run the algorithm 1000 times with random initial variables within the search region, and use the maximum result to approximate $d_i(L')$. In Fig. 4.1, approximated $d_i(L')$ versus L' for TH $= 10^{-4}$, $L_p = 4K - 1$, and $K = 256$ are depicted. It is observed that $d_i(L')$ for $i = 1, 2$ are sufficiently small when $L' \geq 3$, that is, $d_1(3) \ll$ TH and $d_2(3) \ll 1$. To verify that the approximation (4.45) is good for $L' = 3$, we increase L' gradually and solve **P8** for each L'. From Table 4.1, it is observed that the resulted objective value almost stops decreasing for $L' = 3$ and above, which verifies that the approximation (4.45) is good for $L' = 3$. Similar results are obtained for other combinations of parameters. Therefore, we set $L' = 3$ in the following simulations.

For the branch and bound algorithm to solve **P8**, we stop the iteration when the differences between the lower bounds of the remaining subsets and the minimum upper bound are within 0.5×10^{-11}. The computation was performed by a MATLAB program running on a desktop computer

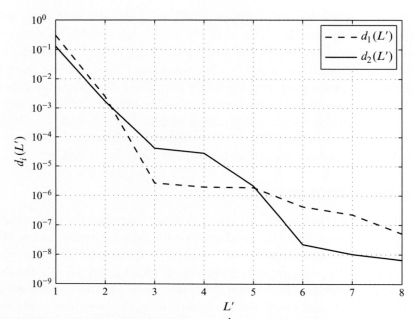

Fig. 4.1 The approximated $d_i(L')$ for TH $= 10^{-4}$, $L_p = 4K - 1$ with $K = 256$, $i = 1, 2$.

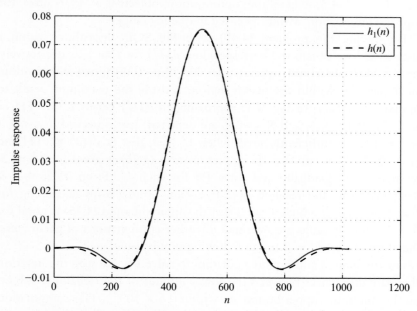

Fig. 4.2 The impulse responses of $h(n)$ and $h_1(n)$ with TH $= 10^{-4}$, $K = 256$, and $L_p = 4K - 1$.

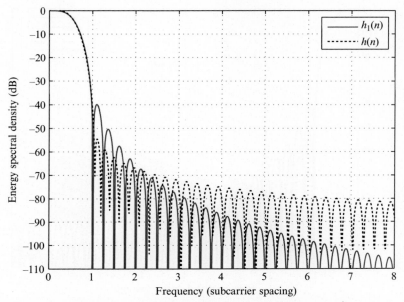

Fig. 4.3 The energy spectral densities of $h(n)$ and $h_1(n)$ with TH $= 10^{-4}$, $K = 256$, and $L_p = 4K - 1$.

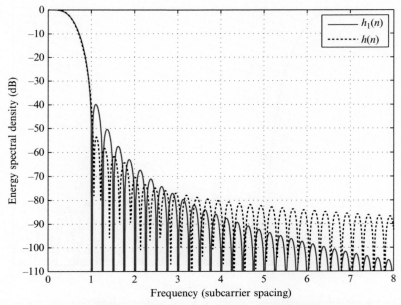

Fig. 4.4 The energy spectral densities of $h_1(n)$ and $h(n)$ with $W(w) = W_1(w)$, TH $= 10^{-4}$, $K = 256$, and $L_p = 4K - 1$.

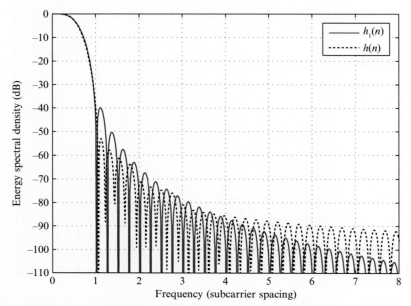

Fig. 4.5 The energy spectral densities of $h_1(n)$ and $h(n)$ with $W(w) = W_2(w)$, TH = 10^{-4}, $K = 256$, and $L_p = 4K - 1$.

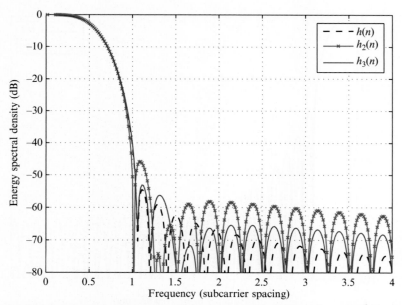

Fig. 4.6 The energy spectral densities of $h(n)$, $h_2(n)$, and $h_3(n)$ with TH = 10^{-4}, $K = 256$, and $L_p = 4K - 1$.

Table 4.1 Objective values of the optimized filters $h(n)$ with TH $= 10^{-4}$, $K = 256$, $L_p = 4K - 1$, and varied L'

Filter	L'	Objective value
$h(n)$	3	2.402×10^{-8} (-76.1943 dB)
	4	2.398×10^{-8} (-76.2015 dB)
	5	2.398×10^{-8} (-76.2015 dB)
	6	2.398×10^{-8} (-76.2015 dB)
	7	2.397×10^{-8} (-76.2033 dB)
	8	2.397×10^{-8} (-76.2033 dB)

equipped with an Intel 3.1 GHz CPU. The convergence time for the branch and bound algorithm with TH $= 10^{-4}$, $L_p = 4K-1$, and $K = 256$ is about 1803 s, which is acceptable considering that the design procedure is needed to be performed only once for typical communication systems.

To verify the performance advantage of the previous optimized filters, we compare them with three types of filters listed as follows.

1. Original frequency sampling filters: Filters obtained in [8], which are denoted by $h_1(n)$. We are interested in this comparison because [12] concludes that it is very hard to find better solution than the filters obtained with the frequency sampling technique.

2. Optimized frequency sampling filters: Optimized filters using the frequency sampling-based technique (optimized with the same objective and constraints as those of **P**1), which are denoted by $h_2(n)$. The frequency sampling-based filter is expressed as [7], such as,

$$h_2(n) = P(0) + 2\sum_{i=1}^{\alpha-1}(-1)^i P(i) \cos\left(\frac{2\pi i}{\alpha K}(n+1)\right), \quad n = 0, 1, \ldots, \alpha K - 2,$$

$$(4.52)$$

where $P(i)$, $i = 0, \ldots, \alpha - 1$ are adjustable coefficients. We substitute $h_2(n)$ into the optimization problem **P**1 and solve **P**1 to obtain the optimized filter. For $L_p = 3K - 1$, α is set to 3, and $P(0) = 1$, $P(2) = \sqrt{1 - P(1)^2}$. For $L_p = 4K - 1$, α is set to 4, and $P(0) = 1$, $P(2) = \sqrt{2}/2$, $P(3) = \sqrt{1 - P(1)^2}$. Therefore, only one parameter, $P(1)$, is required to be optimized for both cases. The optimized frequency sampling filter is obtained by optimizing $P(1)$ with the same objective and constraints as those of the optimization problem **P**1.

3. Optimized windowing-based filters: Optimized filters using the windowing-based technique (optimized with the same objective and constraints as those of **P**1), which are denoted by $h_3(l)$. The windowing-based filter is expressed as [11], such as,

$$h_3(n) = w(n)h_c(n), \tag{4.53}$$

where $h_c(n)$ is given as

$$h_c(n) = \frac{\sin\left[w_c(n - (L_p - 1)/2)\right]}{\pi(n - (L_p - 1)/2)}, \quad n = 0, 1, \ldots, L_p - 1, \tag{4.54}$$

$$w(n) = \sum_{i=0}^{3} (-1)^i A_i \cos\left(\frac{2\pi i n}{L_p - 1}\right). \tag{4.55}$$

The optimized windowing-based filter is obtained by optimizing the cut-off frequency w_c and four weights $A(i)$, $i = 0, 1, 2, 3$, with the same objective and constraints as those of **P**1.

The impulse responses of $h(n)$ and $h_1(n)$ with TH $= 10^{-4}$, $K = 256$, and $L_p = 4K - 1$ are presented in Fig. 4.2 as an example. Noticeable differences between $h(n)$ and the corresponding filter obtained with the frequency sampling technique in [8] can be observed from the figure. Fig. 4.3 shows the energy spectral densities of $h(n)$ and $h_1(n)$. It is observed from Fig. 4.3 that the sidelobes of $h(n)$ within the normalized frequency range between subcarrier 1 and subcarrier 2, which dominate the overall stopband energy, are significantly lower than those of $h_1(n)$.

For some applications, it is needed to lower the sidelobes further in certain frequency range. To satisfy this requirement, one can adjust the weights $W(w)$ in the optimization problem. In Fig. 4.4, we obtain the optimized filter $h(n)$ with TH $= 10^{-4}$, $K = 256$, $L_p = 4K - 1$, and

$$W(w) = W_1(w) = \begin{cases} 1, & w_0 \leq w < 2w_0, \\ 4, & 2w_0 \leq w \leq \pi. \end{cases} \tag{4.56}$$

It is shown that the sidelobes of $h(n)$ with $W_1(w)$ within subcarrier 1 and subcarrier 3 are lower than those of $h_1(n)$ with TH $= 10^{-4}$, $K = 256$, and $L_p = 4K - 1$. In Fig. 4.5, we obtain the optimized filter $h(n)$ with

$$W(w) = W_2(w) = \begin{cases} 1, & w_0 \leq w < 2w_0, \\ 4, & 2w_0 \leq w < 3w_0, \\ 15, & 3w_0 \leq w \leq \pi. \end{cases} \tag{4.57}$$

It is shown that the sidelobes of $h(n)$ with $W_2(w)$ within the subcarrier 1 and subcarrier 4 are lower than those of $h_1(n)$ with TH $= 10^{-4}$, $K = 256$, and $L_p = 4K - 1$. These examples show that the sidelobes of the designed filters within specified frequency ranges can be delicately controlled by adjusting the weights in the optimization formulation, so that it is lower than those of the frequency sampling technique base filters.

In Fig. 4.6, we solve the optimization problem of **P1** using the frequency sampling and windowing-based techniques, and compare the resulted filters for TH $= 10^{-4}$, $K = 256$, and $L_p = 4K - 1$. The objective value of $h_2(n)$ and $h_3(n)$ is 1.2502×10^{-7} (-69.0302 dB) and 4.028×10^{-8} (-73.9487 dB), respectively, which are both greater than the objective value of $h(n)$, 2.402×10^{-8} (-76.1941 dB). The reason that $h(n)$ achieves the best objective is: the constraint approximation is reasonable for the direct optimization of all filter coefficients, while the other two methods only optimize few parameters of the filter.

In Table 4.2, the objective values of $h(n)$ with different combinations of parameters are compared with those of the original frequency sampling filters $h_1(n)$, the optimized frequency sampling filters $h_2(n)$ and the optimized windowing-based filters $h_3(n)$, respectively. It is observed that the objective values (which represent the overall stopband energy) of $h(n)$ are significantly lower than those of the corresponding filters for comparison. In average, the objective values of $h(n)$ are 12.0341 dB lower than the original frequency sampling filters $h_1(n)$, 7.9751 dB lower than the optimized frequency sampling filters $h_2(n)$, and 3.3970 dB lower than the optimized windowing-based filters $h_3(n)$. Particularly, when TH and L_p are larger (TH $= 10^{-3}$ and $L_p = 4K - 1$), the performance gain of $h(n)$ over other three filters is larger: in average, the objective values of $h(n)$ are 17.6170 dB lower than the original frequency sampling filters $h_1(n)$, 13.3343 dB lower than the optimized frequency sampling filters $h_2(n)$, and 7.4822 dB lower than the optimized windowing-based filters $h_3(n)$.

To verify that the optimized filters $h(n)$, $h_2(n)$, and $h_3(n)$ conform to the ISI/ICI requirements, we simulate a pair of directly connected OQAM/FBMC transmitter and receiver with the designed filters and compute the mean squared error (MSE) between the transmitted symbols and received symbols. The transmitted signals are 4QAM modulated, and TH $= 10^{-4}$, $K = 256$, $L_p = 4K - 1$, or $3K - 1$. The MSEs with $h_1(n)$, $h_2(n)$, and $h_3(n)$ are also presented as comparison. Table 4.3 shows that the MSEs with the optimized filters are all successfully constrained within the

Table 4.2 Objective values of the optimized filters $h(n)$ and filters $h_1(n)$, $h_2(n)$, $h_3(n)$

Parameters	Filter	Objective value
$TH = 10^{-4}$ $K = 64$ $L_p = 3K - 1$	$h(n)$ $h_1(n)$ $h_2(n)$ $h_3(n)$	1.19320×10^{-6} (-59.2329 dB) 1.005929×10^{-5} (-49.9743 dB) 4.18962×10^{-6} (-53.7783 dB) 1.80235×10^{-6} (-57.4416 dB)
$TH = 10^{-4}$ $K = 64$ $L_p = 4K - 1$	$h(l)$ $h_1(n)$ $h_2(n)$ $h_3(n)$	9.963×10^{-8} (-70.0161 dB) 1.34760×10^{-6} (-58.7044 dB) 5.0528×10^{-7} (-62.9647 dB) 1.7139×10^{-7} (-67.6601 dB)
$TH = 10^{-4}$ $K = 256$ $L_p = 3K - 1$	$h(n)$ $h_1(n)$ $h_2(n)$ $h_3(n)$	2.8281×10^{-7} (-65.4851 dB) 2.51483×10^{-6} (-55.9949 dB) 1.03242×10^{-6} (-59.8614 dB) 4.1726×10^{-7} (-63.7959 dB)
$TH = 10^{-4}$ $K = 256$ $L_p = 4K - 1$	$h(n)$ $h_1(n)$ $h_2(n)$ $h_3(n)$	2.402×10^{-8} (-76.1943 dB) 3.3690×10^{-7} (-64.7250 dB) 1.2502×10^{-7} (-69.0302 dB) 4.028×10^{-8} (-73.9491 dB)
$TH = 10^{-3}$ $K = 64$ $L_p = 3K - 1$	$h(n)$ $h_1(n)$ $h_2(n)$ $h_3(n)$	1.09290×10^{-6} (-59.6142 dB) 1.005929×10^{-5} (-49.9743 dB) 4.18962×10^{-6} (-53.7783 dB) 1.80235×10^{-6} (-57.4416 dB)
$TH = 10^{-3}$ $K = 64$ $L_p = 4K - 1$	$h(n)$ $h_1(n)$ $h_2(n)$ $h_3(n)$	2.339×10^{-8} (-76.3097 dB) 1.34760×10^{-6} (-58.7044 dB) 5.0528×10^{-7} (-62.9647 dB) 1.3472×10^{-7} (-68.7057 dB)
$TH = 10^{-3}$ $K = 256$ $L_p = 3K - 1$	$h(n)$ $h_1(n)$ $h_2(n)$ $h_3(n)$	2.5917×10^{-7} (-65.8642 dB) 2.51483×10^{-6} (-55.9949 dB) 1.03242×10^{-6} (-59.8614 dB) 4.0676×10^{-7} (-63.9066 dB)
$TH = 10^{-3}$ $K = 256$ $L_p = 4K - 1$	$h(n)$ $h_1(n)$ $h_2(n)$ $h_3(n)$	5.816×10^{-9} (-82.3538 dB) 3.3690×10^{-7} (-64.7250 dB) 1.2502×10^{-7} (-69.0302 dB) 3.167×10^{-8} (-74.9935 dB)

predefined $TH = 10^{-4}$, thus it can be concluded that the optimized filters satisfy the NPR property. Similar simulations for the optimized filters with other combinations of parameters are also performed. The results also show that the MSEs with the optimized filters are all successfully constrained within the predefined TH.

Table 4.3 MSE between the transmitted symbols and the received symbols with TH = 10^{-4} and $K = 256$

Parameters	Filter	MSE (real part)	MSE (imaginary part)
TH = 10^{-4} $K = 256$ $L_p = 3K - 1$	$h(n)$	9.9759×10^{-5}	1.0095×10^{-4}
	$h_1(n)$	4.5362×10^{-5}	4.6218×10^{-5}
	$h_2(n)$	6.5610×10^{-5}	6.5031×10^{-5}
	$h_3(n)$	9.9804×10^{-5}	9.9989×10^{-5}
TH = 10^{-4} $K = 256$ $L_p = 4K - 1$	$h(n)$	1.0187×10^{-4}	1.0003×10^{-4}
	$h_1(n)$	3.0172×10^{-7}	3.0255×10^{-7}
	$h_2(n)$	1.6159×10^{-6}	1.6304×10^{-6}
	$h_3(n)$	1.0160×10^{-4}	9.9791×10^{-5}

4.5 SUMMARY

In this chapter, we presented the problem formulation of direct optimization of the filter coefficients to both minimizing the stopband energy and constraining the ISI/ICI for OQAM/FBMC communication systems. We then introduced the branch and bound algorithm and presented a method to dramatically reduce the number of unknowns of the optimization problem through approximation of the constraints. Simulation results show that the optimized filters obtained with the variable reduction method achieve significantly lower stopband energy than those with the frequency sampling and windowing-based techniques.

REFERENCES

[1] Vaidyanathan PP. Multirate systems and filter banks. Englewood Cliffs, NJ: Prentice-Hall; 1993.
[2] Cvetkovic Z, Vetterli M. Tight Weyl-Heisenberg frames in $\ell^2(\mathbf{Z})$. IEEE Trans Signal Process 1998;46(5):1256–9.
[3] Beaulieu FD, Champagne B. Multicarrier modulation using perfect reconstruction DFT filter bank transceivers. In: Fifth international conference on information, communications and signal processing; 2005.
[4] Scaglione A, Giannakis GB, Barbarossa S. Redundant filterbank precoders and equalizers. I. Unification and optimal designs. IEEE Trans Signal Process 1999;47(7):1988–2006.
[5] Phoong SM, Chang Y, Chen C. DFT-modulated filterbank transceivers for multipath fading channels. IEEE Trans Signal Process 2005;53(1):182–92.
[6] Farhang-Boroujeny B, Kempter R. Multicarrier communication techniques for spectrum sensing and communication in cognitive radios. IEEE Commun Mag 2008;46(4):80–5.
[7] Viholainen A, Ihalainen T, Stitz TH, Renfors M, Bellanger M. Prototype filter design for filter bank based multicarrier transmission. In: 17th European signal processing conference; 2009.

[8] Mirabbasi S, Martin K. Overlapped complex-modulated transmultiplexer filters with simplified design and superior stopbands. IEEE Trans Circuits Syst II Analog Digit Signal Process 2003;50(8):456–69.

[9] Bellanger MG. Specification and design of a prototype filter for filter bank based multicarrier transmission. In: IEEE international conference on acoustics, speech, and signal processing; 2001.

[10] Cruz-Roldan F, Heneghan C, Saez-Landete JB, Blanco-Velasco M, Amo-Lopez P. Multi-objective optimisation technique to design digital filters for modulated multi-rate systems. Electron Lett 2008;44(13):827–8.

[11] Martin-Martin P, Bregovic R, Martin-Marcos A, Cruz-Roldan F, Saramaki T. A generalized window approach for designing transmultiplexers. IEEE Trans Circuits Syst I Regul Pap 2008;55(9):2696–706.

[12] Viholainen A, Bellanger M, Huchard M. Prototype filter and structure optimization, 2009. Available from: http://www.ict-phydyas.org/delivrables/PHYDYAS-D5-1.pdf/view.

[13] Lai X. Optimal design of nonlinear-phase FIR filters with prescribed phase error. IEEE Trans Signal Process 2009;57(9):3399–410.

[14] Davidson T. Enriching the art of FIR filter design via convex optimization. IEEE Signal Process Mag 2010;27(3):89–101.

[15] Chen D, Qu D, Jiang T. Prototype filter optimization to minimize stopband energy with NPR constraint for filter bank multicarrier modulation systems. IEEE Trans Signal Process 2013;61(1):159–69.

[16] Androulakis IP, Maranas CD, Floudas CA. αBB: a global optimization method for general constrained nonconvex problems. J Glob Optim 1995;7(4):337–43.

[17] Adjiman CS, Dallwig S, Floudas CA, Neumaier A. A global optimization method, αBB, for general twice-differentiable constrained NLPs—I. Theoretical advances. Comput Chem Eng 1998;22(9):1137–58.

[18] Adjiman CS, Androulakis IP, Floudas CA. A global optimization method, αBB, for general twice-differentiable constrained NLPs—II. Implementation and computational results. Comput Chem Eng 1998;22(9):1159–79.

[19] Hu J, Wu Z, McCann H, Davis LE, Xie C. Sequential quadratic programming method for solution of electromagnetic inverse problems. IEEE Trans Antennas Propag 2005;53(8):2680–7.

CHAPTER 5

Peak-to-Average Power Ratio

Some major drawbacks still remain unsolved in the design of offset quadrature amplitude modulation-based filter bank multicarrier (OQAM/FBMC) wireless communication systems, which prevent the OQAM/FBMC technique from being implemented in practice. One of the main drawbacks is the large peak-to-average power ratio (PAPR) of the transmitted OQAM/FBMC signals, which is defined as the ratio of the peak power to the average power of the transmitted OQAM/FBMC signals [1]. As is well known, high-power amplifiers (HPAs) are usually utilized in wireless communication systems to amplify the power of transmitted signals to overcome the propagation path loss. Since the HPA employed in OQAM/FBMC systems has limited linear region, the OQAM/FBMC signals with high PAPR will be seriously clipped and nonlinear distortion will be introduced, resulting in serious degradation of the bit error rate (BER) performance [2]. Moreover, the high PAPR leads to the out-of-band radiation, which causes serious adjacent channel interferences. Furthermore, the nonlinear characteristic of the HPA is very sensitive to the variation in signal amplitudes. If the HPA is not operated in the linear region with large power back-off, it is impossible to keep the out-of-band power below the specified limits. This situation leads to low energy efficiency and very expensive transmitters. Therefore, it is of great importance to reduce the PAPR in OQAM/FBMC systems [3].

Recently, various methods have been proposed to reduce the PAPR for OFDM systems in the literature, such as clipping [4, 5], coding schemes [6, 7], nonlinear companding transforms [8–10], partial transmit sequence (PTS) [11, 12], selective mapping (SLM) [13], active constellation extension [14], tone reservation [2, 15], and tone injection [16]. These PAPR reduction techniques can efficiently reduce the PAPR in OFDM systems; however, they cannot be directly utilized in OQAM/FBMC systems because the time-domain OQAM/FBMC signals are overlapped. We will summarize the PAPR reduction methods for OQAM/FBMC systems in

OQAM/FBMC for Future Wireless Communications
http://dx.doi.org/10.1016/B978-0-12-813557-0.00005-X

© 2018 Elsevier Ltd.
All rights reserved.

detail, that is, the alternative signal (AS) [17], multiblock joint optimization (MBJO) [18], and segmental PTS [19]. Moreover, we will also discuss the joint PAPR reduction and sidelobe suppression for noncontiguous OQAM/FBMC-based cognitive radio (CR) networks [20].

The rest of this chapter is organized as follows. Section 5.1 describes the PAPR distribution of OQAM/FBMC signals. Section 5.2 discusses the OQAM/FBMC effect on power amplifiers. The AS method for PAPR reduction is presented in Section 5.3. The MBJO for PAPR reduction is discussed in Section 5.4. The segmental PTS scheme for PAPR reduction is discussed in Section 5.5. Section 5.6 discusses the joint PAPR reduction and sidelobe suppression for noncontinuous OQAM/FBMC signals. Finally, summary is made in Section 5.7.

5.1 PAPR DISTRIBUTION OF OQAM/FBMC SIGNALS

This section discusses the PAPR distribution of OQAM/FBMC signals in detail.

5.1.1 PAPR of OQAM/FBMC Signals

Fig. 5.1 shows the transmitter of OQAM/FBMC systems. At the transmitter, the complex input symbols are written as

$$x_k(m) = a_k(m) + j \times b_k(m), \quad 0 \le k \le K-1, \quad 0 \le m \le M-1, \quad (5.1)$$

where $a_k(m)$ and $b_k(m)$ are the real and imaginary parts of the mth symbol on the kth subcarrier, respectively. The mth symbols on all K subcarriers form a data block $\mathbf{x}(m) = [x_0(m), x_1(m), \ldots, x_{K-1}(m)]^T$. The real and imaginary components of the symbols are staggered in time domain by $T/2$, where T is the symbol period (data block period). The symbols are then passed through a bank of transmission filters and are modulated with K subcarrier modulators whose carrier frequencies are $1/T$-spaced apart. Then, the OQAM/FBMC modulated signal of a burst of M data block is

$$s(t) = \sum_{m=0}^{M-1} s_m(t) = \sum_{k=0}^{K-1} s^k(t) = \sum_{k=0}^{K-1}\sum_{m=0}^{M-1} s_m^k(t)$$

$$= \sum_{k=0}^{K-1}\sum_{m=0}^{M-1} [a_k(m)h(t-mT) + jb_k(m)h\left(t-mT-\frac{T}{2}\right)]e^{jk\varphi_t}, \quad (5.2)$$

$$0 \le t \le \left(M+\alpha-\frac{1}{2}\right)T,$$

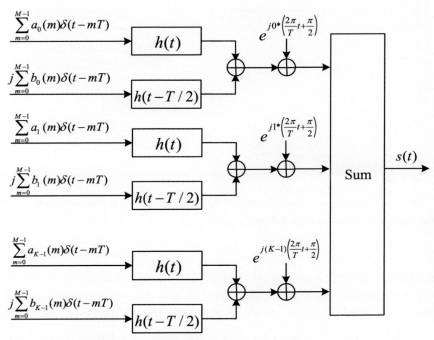

Fig. 5.1 The transmitter of OQAM/FBMC systems.

where $s_m^k(t)$ is the time-domain OQAM/FBMC signal for symbol x_m^k, $h(t)$ is the impulse response of the prototype filter and $\varphi_t = \frac{2\pi t}{T} + \frac{\pi}{2}$. It is assumed that $h(t)$ has finite length αT in time domain. Apparently, the signals of adjacent data blocks overlap with each other due to the fact that, the impulse response of prototype filters in OQAM/FBMC systems has longer time duration than T, and the real and imaginary parts of input symbols are time staggered. Finally, the OQAM/FBMC signal $s(t)$ is modulated to the RF band in order to be transmitted.

In the conventional OFDM systems, the OFDM signals are obtained by taking the inverse fast-Fourier transform operation over the K QAM symbols, and the length of each OFDM signal is T. Since the OFDM symbol rate is $1/T$, there is no overlap between any adjacent OFDM symbols, and the PAPR of each OFDM symbol is defined as the ratio of the peak power to the average power. As shown in Fig. 5.2, the time-domain OQAM/FBMC signals are overlapped with adjacent signals, which are different from the OFDM system. Due to the special signal structure,

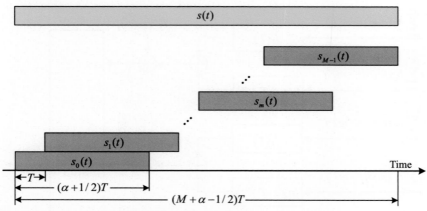

Fig. 5.2 Structure of the OQAM/FBMC signals.

the PAPR definition for the OQAM/FBMC signals needs to be modified. OQAM/FBMC signal $s(t)$ is first divided into $M + \alpha$ intervals, each of which has length T. Then, the PAPR for $s(t)$ in the pth interval is defined as

$$PAPR_p = 10\log_{10}\frac{\max_{pT\leq t\leq(p+1)T}|s(t)|^2}{P_{\text{ave}}}, \quad p = 0, 1, \ldots, M + \alpha - 1,$$

(5.3)

where P_{ave} is the average power of $s(t)$.

5.1.2 Distribution of PAPR

This subsection introduces the distribution of the PAPR of OQAM/FBMC signals. To simplify Eq. (5.2), denote $X_k(m) = a_k(\frac{m}{2})$ if m is even, and $X_k(m) = b_k(\frac{m-1}{2})$ if m is odd. Then, Eq. (5.2) can be rewritten as

$$
\begin{aligned}
s(t) &= \sum_{k=0}^{K-1} s^k(t) = \sum_{k=0}^{K-1}\sum_{m=0}^{M-1} s_m^k(t) \\
&= \sum_{k=0}^{K-1}\sum_{m=0}^{M-1} X_k(m)\underbrace{h\left(t - m\frac{T}{2}\right)e^{j\frac{2\pi k}{T}t}e^{j\phi_m^k}}_{h_m^k(t)},
\end{aligned}
$$

(5.4)

where $\phi_m^k = \frac{\pi}{2}(m + k) - \pi mk$.

Thus, the time-domain signal of the kth subcarrier can be expressed as

$$s^k(t) = \sum_{m=0}^{M-1} s_m^k(t) = \sum_{m=0}^{M-1} X_k(m)h\left(t - m\frac{T}{2}\right) e^{\frac{j2\pi kt}{T}} e^{j\phi_m^k}. \quad (5.5)$$

Suppose that the input data streams $X_k(m)$ are statistically independent and identically distributed (i.i.d.), and the variance of $X_k(m)$ is σ_x^2. Then, according to Eq. (5.5), the expectation and variance of $s^k(t)$ are

$$E[s^k(t)] = 0,$$

$$\sigma_k^2 = E[s^k(t)s^{k\star}(t)] = \sigma_x^2 \sum_{m=0}^{M-1} h\left(t - m\frac{T}{2}\right)^2, \quad (5.6)$$

respectively. Obviously, $E[s^k(t)]$ is uncorrelated with σ_k^2 and k, and $s(t) = \sum_{k=0}^{K-1} s^k(t)$. Therefore, based on the central limit theorem, when K is considerably large, the real part and imaginary part of $s(t)$ approaches Gaussian distribution with zero mean and variance $\sigma_s^2 = K\sigma_k^2/2$. In other words, $|s(t)|^2$ follows a central chi-squared distribution [1]. Supposing $Y = |s(t)|^2$, the probability density function (PDF) of Y can be expressed as

$$P_Y(y) = \frac{1}{2\sigma_s^2} e^{-\frac{y}{2\sigma_s^2}}. \quad (5.7)$$

Then, supposing $Z = |s(t)|^2/E[|s(t)|^2]$, the PDF of Z is

$$P_Z(z) = 2\sigma_x^2 P_Y(2\sigma_x^2 z) = \alpha_t e^{-\alpha_t z}, \quad (5.8)$$

where $\alpha_t = \frac{2}{K\sum_{m=0}^{M-1} h(t-m\frac{T}{2})^2}$.

Thus, the distribution of z can be written as

$$Prob(z \le \gamma) = 1 - e^{-\alpha_t \gamma}. \quad (5.9)$$

Therefore, the distribution of PAPR can be written as

$$Prob(PAPR \le \gamma) = Prob\left(\cap_{i=pT}^{(p+1)T-1} |s_0(t)|^2 \le \gamma\right)$$

$$= \prod_{i=pT}^{(p+1)T-1} Prob(|s_0(t)|^2 \le \gamma) = \prod_{i=pT}^{(p+1)T-1} (1 - e^{-\alpha_t \gamma}). \quad (5.10)$$

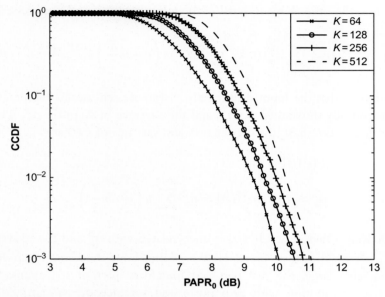

Fig. 5.3 PAPRs of OQAM/FBMC signals with different K.

The complementary cumulative density function (CCDF) of the PAPR systems can be written as

$$Prob(PAPR > \gamma) = 1 - \prod_{i=pT}^{(p+1)T-1} (1 - e^{-\alpha_i \gamma}). \qquad (5.11)$$

Fig. 5.3 shows the CCDF of OQAM/FBMC signals with $K = 64$, 128, 256, and 512 subcarriers. From Fig. 5.3, it is obvious that the PAPR of OQAM/FBMC signals is high, and the PAPR increases when the number of subcarriers increases.

5.2 PAPR EFFECT ON POWER AMPLIFIERS

In practical wireless communication systems, HPAs are usually utilized to amplify the power of transmitted signals to overcome the propagation path loss. Since the HPA employed in OQAM/FBMC systems has limited linear region, the OQAM/FBMC signals with high PAPR will be seriously clipped and nonlinear distortion will be introduced, resulting in serious degradation of the BER performance. Moreover, the nonlinear character-istic of the HPA is very sensitive to the variation in signal amplitudes. If the HPA is not operated in the linear region with large power back–off, it is impossible to keep the out–of–band power below the specified limits. This situation leads to low energy efficiency and very expensive transmitters.

In wireless communication systems, the most widely used HPA is the solid–state power amplifier (SSPA) [21]. The relationship between the input signal $s(t)$ and the output signal $y(t)$ of the SSPA is illustrated in Fig. 5.4 and can be written as

$$y(t) = \frac{s(t)}{\left(1 + \left(\frac{|s(t)|}{A_o}\right)^{2p}\right)^{\frac{1}{2p}}}, \tag{5.12}$$

where A_o denotes the maximum output amplitude of SSPA and p is the smoothness factor of the SSPA from the linear region to the saturation region. The operating point of the SSPA is set by choosing the input back-off (IBO), whichi is defined as

$$IBO = 10 \log_{10} \frac{A_{sat}^2}{P_{in}}, \tag{5.13}$$

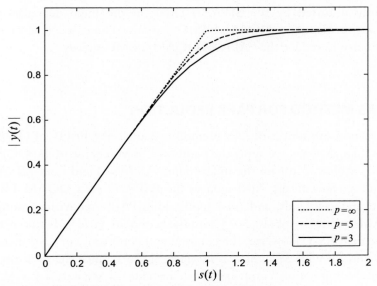

Fig. 5.4 Amplitude-amplitude characteristic of the SSPA.

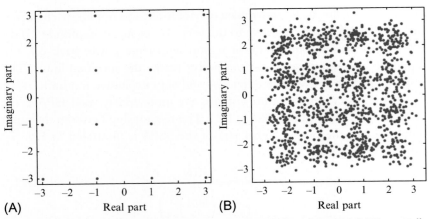

Fig. 5.5 The effects of SSPA on OQAM/FBMC signals. (A) The original 16-QAM constellations. (B) Effect of HPA.

where A_{sat} denotes the maximum input amplitude of the SSPA and P_{in} denotes the average power of OQAM/FBMC signals.

Fig. 5.5 shows the effects of SSPA on OQAM/FBMC signals with $p = 3$ and $A_0 = 3$. It is obvious that the OQAM/FBMC symbols are seriously distorted by the SSPA, resulting in the huge degradation of BER performance in OQAM/FBMC systems. Therefore, it is of great importance to reduce the PAPR of OQAM/FBMC signals.

5.3 AS METHOD FOR PAPR REDUCTION

This section introduces the AS method to reduce the PAPR of OQAM/FBMC signals. First, we apply the traditional SLM scheme to the OQAM/FBMC systems to obtain the independent AS (AS-I) and joint AS (AS-J) algorithms. Specifically, AS-I reduces the PAPR of each OQAM/FBMC symbol independently, and AS-J applies joint PAPR reduction among M OQAM/FBMC symbols. AS-J intuitively should yield a better performance than AS-I. However, the computation complexity of AS-J increases exponentially with M, which is impractical. To balance the performance and the computation complexity, we introduce a sequential AS (AS-S) algorithm, which adopts a sequential optimization procedure over time with the computation complexity increasing linearly with M.

5.3.1 AS-I Algorithm

Inspired by the SLM method, the AS–I algorithm reduces the PAPR by optimally choosing one phase rotation vector from a given set for each OQAM/FBMC symbol. Over different OQAM/FBMC symbols, the phase rotation vectors might be different. Denote the set of candidate phase rotation vectors as

$$\mathcal{B} = \left\{ \boldsymbol{b}^0, \boldsymbol{b}^1, \ldots, \boldsymbol{b}^{U-1} \right\}, \tag{5.14}$$

where U is the size of $\mathcal{B}$, and $\boldsymbol{b}^u, 0 \le u \le U - 1$, is the uth phase rotation vector as

$$\boldsymbol{b}^u = \left[b^{0,u}, b^{1,u}, \ldots, b^{K-1,u} \right]^T, \tag{5.15}$$

with $b^{k,u} = e^{j\frac{2\pi i}{W}}, i = 0, 1, \ldots, W - 1$. We adopt $W = 2$ for simplicity. For convenience, denote $\boldsymbol{b}_m^u = \left[b_m^{0,u}, b_m^{1,u}, \ldots, b_m^{K-1,u} \right]^T$ as the phase rotation vector used by the mth OQAM/FBMC signal $s_m(t)$. Usually, $\mathcal{B}$ is assumed to be known at both the transmitter and the receiver.

After $s_m^k(t)$ is generated as in Eq. (5.2), AS–I generates $\tilde{s}_m^k(t)$ by multiplying the corresponding element in the selected phase rotation vector, that is,

$$\tilde{s}_m^k(t) = s_m^k(t) b_m^{k,u}. \tag{5.16}$$

Then, the new OQAM/FBMC signal $\tilde{s}_m(t)$ is expressed as

$$\tilde{s}_m(t) = \sum_{k=0}^{K-1} s_m^k(t) b_m^{k,u}. \tag{5.17}$$

Thus, PAPR reduction problem with the AS–I algorithm for the mth OQAM/FBMC signal $s_m(t), m = 0, 1, \ldots, M - 1$, can be formulated as

$$(\text{P1}): \min_{\boldsymbol{b}_m^u} \max_{mT \le t \le (m+A+\frac{1}{2})T} \left| \sum_{k=0}^{K-1} s_m^k(t) b_m^{k,u} \right|^2 \tag{5.18}$$

$$\text{subject to: } \boldsymbol{b}_m^u \in \mathcal{B}.$$

Note that we adopt the peak power as the design metric throughout this correspondence. This is because the PAPR reduction should come from the peak power reduction rather than the average power increasing [13]. Given the finite dimensionality of $\mathcal{B}$, an exhaustive search is adopted here to search the optimal $\boldsymbol{b}_m^u$. For each $s_m(t)$, the complexity of searching the optimal $\boldsymbol{b}_m^u$

is on the order of $\mathcal{O}(U)$ (i.e., for each $s_m(t)$), we take U searches. Thus, the complexity for all $s_m(t)$, $m = 0, 1, \ldots, M - 1$, is on the order of $\mathcal{O}(UM)$. Moreover, $\log_2(U)$ bits are needed for side information transmission of each OQAM/FBMC symbol, and thus $M \log_2(U)$ bits for all the M symbols.

The AS-I algorithm does not perform well enough, since it ignores the structure of the OQAM/FBMC signals (i.e., the correlation among adjacent OQAM/FBMC symbols), while reducing the PAPR of $s_m(t)$ independently is strictly suboptimal. To improve the PAPR reduction performance, the AS-J algorithm is introduced in the next subsection to fully explore the intersymbol correlations.

5.3.2 AS-J Algorithm

For each OQAM/FBMC signal $s_m(t)$, the AS-J algorithm firstly chooses one phase rotation vector from the given $\mathcal{B}$; then it applies a joint PAPR reduction scheme among all the M OQAM/FBMC symbols.

Similarly, after $\tilde{s}_m^k(t)$ is generated in the AS-I algorithm, the PAPR reduction problem could be formulated as

$$(\textbf{P2}): \quad \min_{\boldsymbol{b}_m^u, \ldots, \boldsymbol{b}_{M-1}^u} \max_{0 \le t \le (M+A-\frac{1}{2})T} \left| \sum_{m=0}^{M-1} \sum_{k=0}^{K-1} s_m^k(t) b_m^{k,u} \right|^2 \tag{5.19}$$
$$\text{subject to: } \boldsymbol{b}_m^u \in \mathcal{B}, \quad m = 0, 1, \ldots, M - 1.$$

It is obvious that the complexity of exhaustive searching to solve Problem (**P2**) is on the order of $\mathcal{O}(U^M)$, which makes the exhaustive search method impractical. Similarly, the number of bits for the side information is equal to $M \log_2(U)$.

It can be seen from the previous two subsections that the AS-I algorithm is simple but performs badly, while the AS-J algorithm performs well but bears high complexity. To balance the PAPR reduction performance and the complexity, the AS-S algorithm is introduced in the next subsection.

5.3.3 AS-S Algorithm

The main idea of the AS-S algorithm is illustrated in Fig. 5.6, which shows that the AS-S algorithm adopts a sequential optimization procedure. In the mth block, by taking into account the previous OQAM/FBMC symbols, that is, $s_0(t), s_1(t), \ldots, s_{m-1}(t)$, one can reduce the peak power of $s_m(t)$. A detailed illustration of the algorithm is described next.

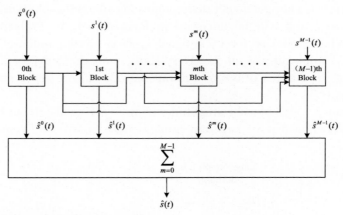

Fig. 5.6 The structure of AS-S algorithm.

In the 0th block, one can multiply $s_0(t)$ by different phase rotation vectors, and choose the one with the minimum peak power, denoted as $\hat{s}_0(t)$. Then, $\hat{s}_0(t)$ is sent to the 1st block to solve the following problem:

$$\min_{\boldsymbol{b}_1^u} \max_{2T \leq t \leq 4T} \left| \hat{s}_0(t) + \sum_{k=0}^{K-1} s_1^k(t) b_1^{k,u} \right|^2 \tag{5.20}$$

$$\text{subject to: } \boldsymbol{b}_1^u \in \mathcal{B}.$$

The optimal phase rotation vector is denoted as $\boldsymbol{b}_1^{u^\star}$, and the new generated signal is cast as

$$\hat{s}_1(t) = \sum_{k=0}^{K-1} s_1^k(t) b_1^{k,u^\star}. \tag{5.21}$$

Next, $\hat{s}_1(t)$ and $\hat{s}_0(t)$ are both sent to the 2nd block to calculate the new signal $\hat{s}^2(t)$. Repeat the previous procedure until the $(M-1)$th block.

Based on the earlier discussion, AS-S is a sequential optimization procedure. Specifically, in the mth block, $m = 1, 2, \ldots, M-1$, the optimization problem could be formulated as

$$(\text{P3}): \min_{\boldsymbol{b}_m^u} \max_{(m+1)T \leq t \leq (m+\Gamma)T} \left| \sum_{l=0}^{m-1} \hat{s}_l(t) + \sum_{k=0}^{K-1} s_m^k(t) b_m^{k,u} \right|^2 \tag{5.22}$$

$$\text{subject to: } \boldsymbol{b}_m^u \in \mathcal{B}.$$

Note that Γ is a key parameter that significantly affects the PAPR reduction performance and will be discussed in Remark 5.1. In Problem (**P3**), the search complexity for each signal $s_m(t)$ is on the order of $\mathcal{O}(U)$, and the complexity for all the M symbols is on the order of $\mathcal{O}(UM)$. Similarly, the number of bits to transmit the side information is also equal to $M \log_2(U)$.

Remark 5.1. We plot the amplitudes of $h(t - mT)$ and $s_m(t)$ in Fig. 5.7, and it is obvious that the large-amplitude samples of $h(t - mT)$ are located

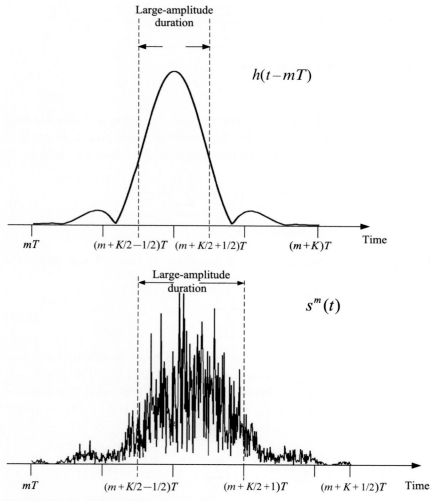

Fig. 5.7 The amplitudes of $h(t - mT)$ and $s_m(t)$, respectively.

within $\{(m+\frac{\alpha}{2}-\frac{1}{2})T \le t \le (m+\frac{\alpha}{2}+\frac{1}{2})T\}$. For $h(t-mT-\frac{T}{2})$, its large-amplitude samples are located within $\{(m+\frac{\alpha}{2})T \le t \le (m+\frac{\alpha}{2}+1)T\}$. According to Eq. (5.2), one could obtain that the large-amplitude samples of $s_m(t)$ are located within $\{(m+\frac{\alpha}{2}-\frac{1}{2})T \le t \le (m+\frac{\alpha}{2}+1)T\}$. Intuitively, to obtain a good PAPR reduction performance, the large-amplitude samples of $s_m(t)$ should be included in the optimization duration $\{(m+1)T \le t \le (m+\Gamma)T\}$ in problem (P3); that is, Γ should satisfy $\Gamma \ge \frac{\alpha}{2}+1$. Furthermore, since $s_m(t)$ only spans over $\{mT \le t \le (m+\alpha+\frac{1}{2})T\}$, it follows that $\Gamma \le (\alpha+\frac{1}{2})$. Based on the earlier discussions, it can be concluded that $\frac{\alpha}{2}+1 \le \Gamma \le (\alpha+\frac{1}{2})$ is a good choice.

Based on the earlier discussion, the AS-S algorithm is summarized as follows:

Step 1: Initialization: $m=1$, multiply $s_0(t)$ by different phase rotation vectors and denote the one with the minimum peak power as $\hat{s}_0(t)$, which is sent to the 1st block.

Step 2: In the mth block, solve Problem (P3), and the new signal is denoted as $\hat{s}_m(t)$, send $\hat{s}_0(t), \hat{s}_1(t), \ldots, \hat{s}_m(t)$ to the next block.

Step 3: Set $m=m+1$, if $m \le M-1$, go to Step 2. Otherwise, calculate $\hat{s}(t) = \sum_{m=0}^{M-1} \hat{s}_m(t)$ and output the value.

5.3.4 Simulation Results

In this subsection, simulations are conducted to evaluate the PAPR reduction performance of the AS-S algorithm. For all the simulations, In total, 10^4 symbols are randomly generated, and the over-sampling factor is adopted as 4. The filter response $h(t)$ is set as

$$h(t) = \alpha_1 \left(c(0) + 2\sum_{i=1}^{\alpha-1}(-1)^i c(i) \cos\left(\frac{2i\pi}{\alpha K}t\right)\right), \qquad (5.23)$$

where α_1 is the normalization factor, $c(i), i=0,1,\ldots,\alpha-1$, are given as

$$c(0) = 1,$$
$$c(1) = 0.97195983,$$
$$c(2) = \frac{1}{\sqrt{2}} \qquad (5.24)$$
$$c(3) = 0.23514695.$$

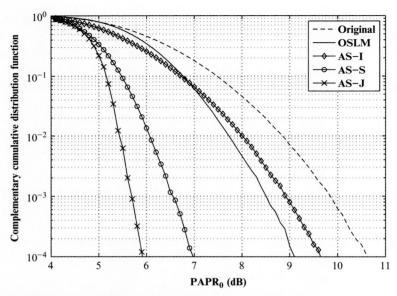

Fig. 5.8 CCDF versus PAPR$_0$ for different schemes, 4-QAM, $M = 4, K = 32, U = 8, \Gamma = 4$.

Fig. 5.8 plots the CCDF curves for all the AS algorithms, where 4-QAM is employed, and choose $K = 32$, $U = 8$, and $\Gamma = 4$. To obtain the CCDF curve for AS-J, one can set M as a small number (i.e., $M = 4$). For comparison, we also plot the CCDF curve for the OSLM method in [22]. It is observed that the PAPR could be reduced by about 1.7 dB with OSLM, by only 1.5 dB with AS-I, and by almost 3.8 dB with AS-S, compared to the original FBMC signals when CCDF $= 10^{-4}$. In addition, AS-J could improve the performance by about 1 dB over AS-S. However, for AS-J, one can run 8^4 searches, which is much more than the 8×4 searches for AS-S.

Fig. 5.9 illustrates the performance with different U values when AS-I and AS-S are deployed, where 16-QAM is adopted, $M = 1000$, $K = 128$, $\Gamma = 4$, and U is selected as $U = 4, 16, 64, 128$, respectively. Since $U^M = U^{1000}$ is a relatively huge number, we do not plot the curves of AS-J. Obviously, compared with the original OQAM/FBMC signal without PAPR reduction at CCDF$= 10^{-4}$, AS-S could provide 2.8, 4.0, 4.5, and 4.7 dB PAPR reduction gains when $U = 4, 16, 64, 128$, respectively. Therefore, increasing U could improve the PAPR reduction

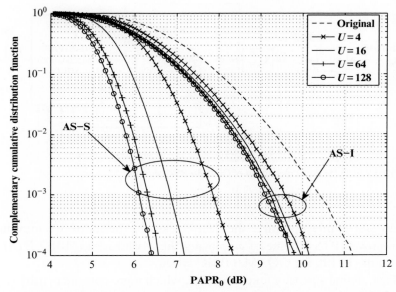

Fig. 5.9 CCDF versus $PAPR_0$ for AS-I and AS-S with different U, 16-QAM, $M = 10^3, K = 128, \Gamma = 4$.

performance significantly for AS-S. However, as U increases, the PAPR reduction performance barely improves for AS-I.

5.4 MBJO FOR PAPR REDUCTION

This section discusses an improved PTS scheme by employing MBJO for the PAPR reduction of OQAM/FBMC signals, called an MBJO-PTS scheme. Unlike existing PAPR reduction schemes of independently optimizing the data blocks, the MBJO-based scheme exploits the overlapping structure of the OQAM/FBMC signal and jointly optimizes multiple data blocks. It is observed that the MBJO-PTS optimization problem can be divided into a sequence of overlapping optimization subproblems. Thus, we introduce a dynamic programming (DP) algorithm to solve the optimization problem without an exhaustive search.

5.4.1 MBJO-PTS Scheme for PAPR Reduction

Fig. 5.10 illustrates the key idea of the MBJO. Unlike the conventional PAPR reduction schemes of optimizing the parameters independently for each data block, an MBJO-based scheme buffers a number of data

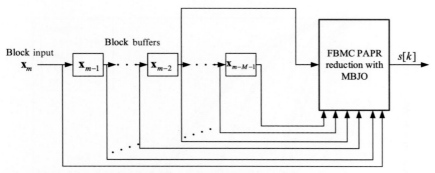

Fig. 5.10 MBJO for PAPR reduction of OQAM/FBMC signals.

blocks and jointly optimizes the data blocks. Furthermore, we introduce an improved PTS scheme of employing the MBJO idea (MBJO-PTS) to reduce the PAPR for OQAM/FBMC systems.

First, define a penalty function $f(P_q)$, which represents the penalty (e.g., signal distortion, error performance loss) to the communication system resulted from the peak power of the qth segment. Apparently, $f(P_q)$ has to be an increasing function of P_q.

The key idea of the MBJO-PTS scheme is to minimize the summed penalty of the signal of all M data blocks, by jointly optimizing the phase factor vectors of M data blocks, that is,

$$\min_{\boldsymbol{\beta}_0,\boldsymbol{\beta}_1,\ldots,\boldsymbol{\beta}_{M-1}} \sum_{q=0}^{M+\alpha-1} f(P_q), \qquad (5.25)$$

$$\text{subject to: } \beta_m^v \in e^{j\frac{2\pi i}{W}}, i = 0, 1, \ldots, W-1, \qquad (5.26)$$

$$v = 1, 2, \ldots, V, \quad m = 0, 1, \ldots, M-1, \qquad (5.27)$$

where the time-domain OQAM/FBMC signals are divided into V sub-blocks, and $\boldsymbol{\beta}_m = [\beta_m^1, \beta_m^2, \ldots, \beta_m^{V-1}]$ is the phase rotation vector of the mth data block. Moreover, solutions of the MBJO-PTS optimization problem will be discussed in the next subsection.

The optimization formulation is called MBJO based on its two features. First, phase factor vectors of multiple data blocks are jointly optimized in the formulation. Joint optimization can deal with the problem of overlapping of data blocks, and even exploit the overlapping for better PAPR reduction. Second, the objective is a sum of penalty of the signal segments covering multiple data blocks. The summed penalty is an overall measurement of

PAPR of the signal, which is more suitable for the signal with overlapping structure than separated value of peak power. Although the OSLM method proposed in [22] also considered the overlapping of data blocks, one cannot call it as an MBJO-based method due to the following two reasons. First, each data block is optimized separately in OSLM. Second, there is no segmentation of signal in OSLM, and the objective of the OSLM optimization is the peak power of the signal in the scope of the current data block, not a summed value with the whole signal taken into account.

5.4.2 Optimal Solutions to the MBJO-PTS Optimization

Since each data block has W^V candidate phase factor vectors, there are W^{VM} possible combinations of phase factor vectors for the OQAM/FBMC signal with M data blocks. Thus, the computational complexity of solving the MBJO-PTS optimization problem by exhaustive search is $O(W^{VM})$, which is unacceptable in practice.

We introduce a DP algorithm to solve the optimization problem in Eq. (5.25) without an exhaustive search. DP is an optimization technique of solving complex problems by breaking them down into simpler steps. It is applicable to problems, which exhibit the characteristics of overlapping subproblems and optimal substructure. The DP algorithm can find the optimal solution for this kind of problems with low computational complexity. In this subsection, the similarity between the mathematical model of the DP algorithm and our optimization problem motivates us to design the DP algorithm for the MBJO-PTS, called MBJO-PTS-DP.

For simplicity of presentation, we first present the MBJO-PTS-DP algorithm for $\alpha = 1$ (i.e., for the case when the signal of a data block only overlaps with the signals of both one preceding and one succeeding data blocks). We then briefly discuss how to modify the algorithm for larger A. To apply the DP technique, a trellis diagram for the optimization problem has to be constructed first, by breaking down the problem into a sequence of serially connected stages (subproblems). Each stage represents one OQAM/FBMC data block and has a number of states that represent the candidate phase factor vectors for the corresponding data block. For the lth state, its corresponding phase factor vector is given as

$$\boldsymbol{\theta}(l) = [e^{j2\pi \frac{mod(l,W)}{W}}, e^{j2\pi \frac{mod(l,W^2)-mod(l,W)}{W^2}}, \dots, e^{j2\pi \frac{l-mod(l,W^{V-1})}{W^V}}], \quad 0 \le l \le W^V - 1,$$
$$(5.28)$$

where $mod(a, b)$ represents the function of a modulo b. Thus, there are M stages and each stage has W^V states for an OQAM/FBMC signal with M

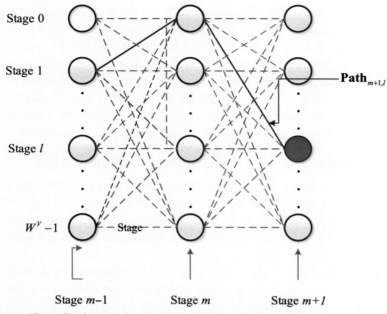

Fig. 5.11 The trellis diagram for the MBJO-PTS-DP algorithm with $\alpha = 1$.

data blocks. As shown in Fig. 5.11, the trellis diagram depicts the stages and their corresponding states. Obviously, a path that travels from the first stage to the last stage in the trellis diagram represents an OQAM/FBMC signal with a particular combination of phase factor vectors. Thus, the aim is to find the path leading to the lowest objective defined by Eq. (5.25).

The DP algorithm locates the path of the lowest objective, without the need for exhaustive searching. The key idea of the DP algorithm is explained as follows: For two paths entering the same state of stage m, their objective up to this stage is calculated and compared, and the path with the greater objective is discarded. The other path with the lower objective is saved and called the survivor path of this state of stage m. The elimination of one of the two paths may be done without compromising the optimality of the trellis search, because any extension of the path with the greater objective beyond stage m will always have a greater objective than the survivor path that is extended along the same path beyond stage m. Basically, the DP algorithm does the following: (1) Among all paths that entering a state of Stage m, the path with the lowest objective is selected as the survivor path for the state. (2) The survivor paths of Stage $(m + 1)$

are selected by comparing the extensions of the survivor paths of Stage m. (3) At the final stage, the survivor paths of all states are compared, and the survivor path with the lowest objective is the path of lowest objective among all paths in the trellis.

Denote the survivor path that ends at the lth state of the mth stage by the following vector:

$$\mathbf{Path}_{m,l} = [\mathbf{Path}^0_{m,l}, \mathbf{Path}^1_{m,l}, \ldots, \mathbf{Path}^m_{m,l}], \quad 0 \le l \le W^V - 1, \quad (5.29)$$

where $\mathbf{Path}^i_{m,l}, 0 \le i \le m$ represent the state of the ith stage on the path of $\mathbf{Path}_{m,l}$. Apparently, $\mathbf{Path}^m_{m,l} = \boldsymbol{\theta}(l)$, and $[\mathbf{Path}^0_{m,l}, \mathbf{Path}^1_{m,l}, \ldots, \mathbf{Path}^{m-1}_{m,l}]$ is the survivor path at State $\mathbf{Path}^{m-1}_{m,l}$ of Stage $m - 1$.

The MBJO–PTS–DP algorithm for the optimization of Eq. (5.29) is presented as follows in detail.

Step 1: Set $m = 0$. Initialize W^V paths at Stage 0 as $\mathbf{Path}_{0,l} = \boldsymbol{\theta}(l)$, for $0 \le l \le W^V - 1$.

Step 2: Denote $C_{m+1}(l', l)$ as the penalty of the OQAM/FBMC signal corresponding to the path that passes through State l' of Stage m and ends at State l of Stage $m + 1$. With the survivor path $\mathbf{Path}_{m,l'}$, the penalty $C_{m+1}(l', l)$ is calculated as

$$C_{m+1}(l', l) = \sum_{q=0}^{M+\alpha-1} f(P_q), \quad \text{for } 0 \le l, \quad l' \le W^V - 1, \quad (5.30)$$

where

$$[\boldsymbol{\beta}_0, \boldsymbol{\beta}_1, \ldots, \boldsymbol{\beta}_m] = \mathbf{Path}_{m,l'}, \quad \boldsymbol{\beta}_{m+1} = \boldsymbol{\theta}(l), \quad (5.31)$$
$$\boldsymbol{\beta}_{m+2} = \cdots = \boldsymbol{\beta}_{M-1} = \mathbf{0}, \quad (5.32)$$

and $\mathbf{0}$ represents zero vector.

Step 3: The path with the lowest $C_{m+1}(l', l)$ is selected as the survivor path at State l of Stage $(m + 1)$, $\mathbf{Path}_{m+1,l}$. Let

$$l'_{m+1,l} = \arg\min_{l'} C_{m+1}(l', l), \quad \text{for } 0 \le l \le W^V - 1. \quad (5.33)$$

Survivor paths at Stage $(m + 1)$ can be updated as

$$\mathbf{Path}_{m+1,l} = [\mathbf{Path}_{m+1,l'_{m+1,l}}, \boldsymbol{\theta}(l)], \quad \text{for } 0 \le l \le W^V - 1. \quad (5.34)$$

Step 4: If $M+1 < M-1$, let $m = m+1$, and go to Step 2. If $m+1 = M-1$, stop the iteration. Denote $C_{M-1}(l)$ as the penalty of the path ends at

State l of the last stage (i.e., Stage $M - 1$),

$$C_{M-1}(l) = \sum_{q=0}^{M+\alpha-1} f(P_q), \quad \text{for } 0 \leq l \leq W^V - 1, \tag{5.35}$$

where

$$[\boldsymbol{\beta}_0, \boldsymbol{\beta}_1, \ldots, \boldsymbol{\beta}_{M-1}] = \mathbf{Path}_{M-1,l}. \tag{5.36}$$

Finally, the path that leads to the minimum penalty is obtained as

$$\mathbf{Path} = \mathbf{Path}_{M-1,l^\star}, \tag{5.37}$$

where

$$l^\star = \arg\min_{l} C_{M-1}(l), \tag{5.38}$$

and the elements of **Path** represent the optimal phase factor vectors for all data blocks.

Moreover, when $\alpha > 1$, the trellis diagram is different from that of $\alpha = 1$ and the DP algorithm can be modified accordingly.

5.4.3 Simulation Results

In this subsection, extensive simulations are conducted to investigate the PAPR reduction performance of the MBJO-PTS scheme. The number of the subcarriers K is set to be 64, and the number of data blocks M is set to be 256 for the OQAM/FBMC systems. Moreover, 4QAM modulation and $L = 4$ over-sampling are adopted in the simulations. For the MBJO-PTS scheme, the phase factors are chosen from the set $e^{j\frac{2\pi i}{2}}, i = 0, 1$, that is, $W = 2$ and $\boldsymbol{\beta}_m^v \in [1, -1]$ for all v and m.

The penalty function $f(P_q)$ in the simulations has the form of

$$f(P_q) = e^{\tilde{\alpha} P_q}. \tag{5.39}$$

This form of penalty function gives more weight to high peak power, so that MBJO-PTS schemes with this form of penalty function can better avoid high peak power than those with a linear penalty function. Fig. 5.12 shows the influence of parameter $\tilde{\alpha}$ in penalty function $f(P_q)$ on the performance of the MBJO-PTS scheme with the DP algorithm. It is observed that $\tilde{\alpha} = 1.0$ (i.e., $f(P_q) = e^{P_q}$) is the best one among all simulated values of $\tilde{\alpha}$. Moreover, the gap between the best and worst performance of PAPR reduction when parameter $\tilde{\alpha}$ is within the range of 0.25–4.0 is about 0.2 dB

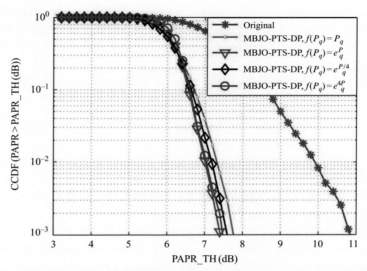

Fig. 5.12 CCDFs of the OQAM/FBMC signal with the MBJO-PTS scheme employing the DP algorithm, different $\tilde{\alpha}$, $V = 4$.

at CCDF of 10^{-3}. Therefore, though the parameter indeed affects the performance of PAPR reduction, it does not make a significant difference in PAPR reduction performance even if $\tilde{\alpha}$ is chosen from a broad range. It is also demonstrated in Fig. 5.12 that the MBJO-PTS scheme using the nonlinear penalty function as presented earlier performs better in PAPR reduction than that using the linear penalty function (i.e., $f(P_q) = P_q$).

5.5 SEGMENTAL PTS SCHEME FOR PAPR REDUCTION

This section introduces a novel segmental PTS scheme, termed an S-PTS scheme for simplicity, to reduce the PAPR in OQAM/FBMC systems. For the S-PTS scheme, the overlapped OQAM/FBMC signals are divided into a number of segments. In each segment, some disjoint subblocks are partitioned, then multiplied with different phase rotation factors. Simulation results verify that the S-PTS scheme could offer better PAPR reduction with lower complexity than the conventional PTS (C-PTS) scheme directly employed in OQAM/FBMC systems.

5.5.1 S-PTS Scheme

For the S-PTS scheme, due to the overlapping structure of OQAM/FBMC signals, one can reduce the peak power of the overlapped OQAM/FBMC signals composed of multiple data blocks, instead of the peak power of each data block independently. Thus, we firstly obtain the overlapped filtered signals, then, divides them into segments and reduces the peak power of each segment.

It is obvious to write the filtered signal on the nth subcarrier as

$$S^k(t) = \sum_{m=0}^{M-1} S_m^k(t) = \sum_{m=0}^{M-1} e^{j\phi_m^k} X_k(m)h(t - m\tau_0), \qquad (5.40)$$

where $\tau_0 = \frac{T}{2}$ is the symbol interval, and $S_m^k(t)$ is the filtered signal on the kth subcarrier of the mth data block with $k = 0, 1, \ldots, K - 1$. Since the filter length is αT and the symbol interval is $\frac{T}{2}$, the lengths of $S_m^k(t)$ and $S^k(t)$ are αT and $(\frac{M}{2} + \alpha)T$, respectively.

As shown in Fig. 5.13, the overlapped filtered signal $S^k(t)$ is divided into several segments in the duration T_s. It is obvious that the number of segments is $D = \frac{(M/2+\alpha)T}{T_s}$. The dth segment ($d = 0, 1, \ldots, D - 1$) is $\mathcal{S}_d = \{S_d^k(t)|k = 0, 1, \ldots, K - 1\}$, where $S_d^k(t)$ is the signal on the kth subcarrier in the dth segment as

$$S_d^k(t) = S^k(t)R_{T_s}(t - dT_s),$$

$$= \sum_{m=0}^{M-1} e^{j\phi_m^k} X_k(m)h(t - m\tau_0)R_{T_s}(t - dT_s), \qquad (5.41)$$

with

$$R_{T_s}(t - dT_s) = \begin{cases} 1, & dT_s \leq t < (d+1)T_s, \\ 0, & \text{else}, \end{cases} \qquad (5.42)$$

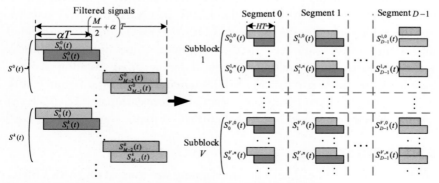

Fig. 5.13 Illustration of segments and subblocks in the S-PTS scheme.

where $R_{T_s}(t - dT_s)$ is a rectangular window function. Indeed, each filtered signal $S_m^k(t)$ is divided into several parts which are distributed in consecutive segments. For simplicity, suppose $T_s = HT, H \geq \frac{\alpha}{2}, H \in \mathbb{Z}$ to ensure that each filtered signal $S_m^k(t)$ is divided into not more than two segments.

Then, K signals in the dth segment $\mathcal{S}_d$ are partitioned into V disjoint subblocks $\mathcal{S}_d^v = \{S_d^{v,k}(t)|k = 0, 1, \ldots, K-1\}$ for $v = 1, 2, \ldots, V$, satisfying

$$S_d^{v,k}(t) = \begin{cases} S_d^k(t), & \frac{K}{V}(v-1) \leq k \leq \frac{K}{V}v - 1, \\ 0, & \text{else.} \end{cases} \tag{5.43}$$

Therefore, when signals $S_d^{v,k}(t), (k = 0, 1, \ldots, K-1)$ are modulated to subcarriers, V signals in the dth segment are obtained as

$$s_d^v(t) = \sum_{k=0}^{K-1} S_d^{v,k}(t)e^{j\frac{2\pi k}{T}t}, \quad v = 1, 2, \ldots, V. \tag{5.44}$$

Finally, the signal $s_d^v(t)$ is multiplied with the phase rotation factor $b_d^v \in \{1, -1\}$, since $\{1, -1\}$ is easily implemented and as good as any other phase sequence in terms of the PAPR reducing capability [23]. The optimal phase factor combination of $s_d^v(t)$ with the minimum PAPR is then selected as

$$\arg\min_{b_d^v} \max_{dT_s \leq t < (d+1)T_s} \left| \sum_{v=1}^{V} b_d^v s_d^v(t) \right|^2 \tag{5.45}$$

$$s.t. \ b_d^v \in \{1, -1\}, \quad d = 0, 1, \ldots, D-1.$$

Since one can firstly obtain the overlapped filtered signals, and then divide the filtered signals into several segments to rotate the phase, all the data blocks are taken into account to select the optimal phase rotation combination. Thus, the S-PTS scheme could achieve good PAPR reduction.

5.5.2 Interference Analysis and Elimination

Due to the dividing operation, some unexpected interferences that impact the signal recovery are introduced. We analyze and eliminate the unexpected interference caused by the divided segment.

Dividing the filtered signal $S^k(t)$ into segments is equivalent to dividing the filter impulse response $h(t)$ into segments. The equivalent filter impulse response is then written as

$$h'(t) = \omega_1 h(t) R_{\tilde{\lambda}_1}(t) + \omega_2 h(t) R_{\tilde{\lambda}_2}(t - \tilde{\lambda}_1), \tag{5.46}$$

where ω_1 and ω_2 are phase rotation factors, and $\tilde{\lambda}_1$ and $\tilde{\lambda}_2$ are the widths of the rectangular windows of the two adjacent segments, respectively, with $\tilde{\lambda}_1 + \tilde{\lambda}_2 = \alpha T$. It is worth noting that the filter would remain unchanged when the phase rotation factors of the two adjacent segments are the same.

Remark. When $S^k(t)$ is divided into segments, $S^k_m(t)$ and $h(t - m\tau_0), (m = 0, 1, \ldots, M - 1)$ are divided at distinct positions. Then, $\{\tilde{\lambda}_1, \tilde{\lambda}_2\}$ are with $2\alpha - 1$ cases, satisfying $\tilde{\lambda}_1 = \frac{T}{2}a, \tilde{\lambda}_2 = \frac{T}{2}(2\alpha - a), (a = 1, 2, \ldots, 2\alpha - 1)$. For example, when $\alpha = 4$, there are seven cases for $\{\tilde{\lambda}_1, \tilde{\lambda}_2\}$: $\{\frac{1}{2}T, \frac{7}{2}T\}$, $\{T, 3T\}$, $\{\frac{3}{2}T, \frac{5}{2}T\}$, $\{2T, 2T\}$, $\{\frac{5}{2}T, \frac{3}{2}T\}$, $\{3T, T\}$, $\{\frac{7}{2}T, \frac{1}{2}T\}$. Note that the case $\{2T, 2T\}$ means that the signal $S^k_m(t)$ is divided from the middle.

Suppose $S^{k_0}_{m_0}(t)$ is divided into two segments, the new intrinsic interference is

$$I^{k_0}_{m_0}{}' = \sum_{\substack{m=0 \\ m \neq m_0}}^{M-1} \sum_{\substack{k=0 \\ k \neq k_0}}^{K-1} X_k(m) \underbrace{\int_{-\infty}^{\infty} h^{k_0}_{m_0}(t) h^{k}_{m}{}^{'*}(t) dt}_{\zeta^{k,k_0}_{m,m_0}{}'}, \qquad (5.47)$$

where $\zeta^{k,k_0}_{m,m_0}{}'$ is the new coefficient for the cases $\{\frac{1}{2}T, \frac{7}{2}T\}$, $\{T, 3T\}$, $\{\frac{3}{2}T, \frac{5}{2}T\}$, and $\{2T, 2T\}$ with $\omega_1 = -1, \omega_2 = 1$ as shown in Table 5.1. It is noted that most real parts of the coefficients are nonzero when $m \neq m_0$ and $k \neq k_0$ for all these cases, then, $I^{k_0}_{m_0}{}'$ is no longer a pure imaginary, and real-valued intrinsic interference is introduced. However, the real-valued interferences are relatively low and most of them are negligible for the cases $\{\frac{1}{2}T, \frac{7}{2}T\}$, $\{T, 3T\}$, and $\{\frac{3}{2}T, \frac{5}{2}T\}$. Meanwhile, the cases $\{\frac{5}{2}T, \frac{3}{2}T\}$, $\{3T, T\}$, and $\{\frac{7}{2}T, \frac{1}{2}T\}$ are symmetric to the cases $\{\frac{3}{2}T, \frac{5}{2}T\}$, $\{T, 3T\}$, and $\{\frac{1}{2}T, \frac{7}{2}T\}$, respectively, thus, they can be omitted in this chapter. Consequently, unexpected real-valued interferences are mainly caused by the filtered signals which are divided from the middle.

To reduce the real-valued interferences, zeros are inserted to the symbols of the case $\{2T, 2T\}$. Thus, we can insert zeros to the m_0th data block, that is, $X_k(m_0) = 0, (k = 0, 1, \ldots, K - 1)$. As an example, the intrinsic interference $I^{k_1}_{m_1}{}'$ that (m_1, k_1) is next to the m_0th data block is written as

$$I^{k_1}_{m_1}{}' = \sum_{\substack{m=0 \\ m \neq m_1}}^{M-1} \sum_{\substack{k=0 \\ k \neq K_1}}^{K-1} X_k(m) \zeta^{k,k_1}_{m,m_1}{}',$$

$$= \sum_{\substack{m=0 \\ m \neq m_1 \\ m \neq m_0}}^{M-1} \sum_{\substack{k=0 \\ k \neq k_1}}^{K-1} X_k(m) \zeta_{m,m_1}^{k,k_1}{}' + \sum_{\substack{k=0 \\ k \neq k_1}}^{K-1} X_k(m_0) \zeta_{m,m_1}^{k,k_1}{}', \qquad (5.48)$$

where $\zeta_{m,m_1}^{k,k_1}{}'$ is the coefficient of the impulse response between the time-frequency points (m_1, k_1) with (m, k) as shown in Table 5.2. From Eq. (5.48) and Table 5.2, one can learn that $I_{m_1}^{k_1}{}'$ is imaginary-valued, since $X_k(m_0) = 0, (k = 0, 1, \ldots, K - 1)$. Thus, the interferences introduced by the case $\{2T, 2T\}$ are completely avoided. Though the real-valued interferences caused by other cases are not avoided, the interferences are relatively small and most of them are negligible, as shown in Table 5.1. Therefore, the real-valued interferences caused by the dividing operation could be decreased.

The distribution of the inserted zero time-frequency points is shown in Fig. 5.14. Since the interval between the real and imaginary parts is $\frac{T}{2}$, the time index p_k of the filtered signal, divided in the middle, is expressed as

Table 5.1 Coefficients of different cases

		$m_0 - 1$	m_0	$m_0 + 1$
$\{\frac{1}{2}T, \frac{7}{2}T\}$	$k_0 - 2$	0	0	0
	$k_0 - 1$	$0.0003 - 0.2497j$	$0.0002 - 0.3184j$	$-0.2501j$
	n_0	$-0.5012j$	0.9994	$0.5013j$
	$k_0 + 1$	$-0.0003 - 0.2497j$	$0.0002 + 0.3184j$	$-0.2501j$
	$k_0 + 2$	0	0	0
$\{T, 3T\}$	$k_0 - 2$	$-0.0027 - 0.0042j$	$0.0012 + 0.0006j$	0
	$k_0 - 1$	$-0.0034 - 0.2522j$	$-0.0005 - 0.3177j$	$0.0003 - 0.2497j$
	n_0	$-0.5075j$	0.9977	$0.5012j$
	$k_0 + 1$	$0.0034 - 0.2522j$	$-0.0005 + 0.3177j$	$-0.0003 - 0.2497j$
	$k_0 + 2$	$0.0027 + 0.0042j$	$0.0012 - 0.0006j$	0
$\{\frac{3}{2}T, \frac{5}{2}T\}$	$k_0 - 2$	$0.0826 - 0.0405j$	$0.0164 + 0.0117j$	$0.0027 - 0.0042j$
	$k_0 - 1$	$-0.1072 - 0.2866j$	$0.0087 - 0.3353j$	$-0.0034 - 0.2522j$
	n_0	$-0.4246j$	0.9728	$0.5075j$
	$k_0 + 1$	$0.1072 - 0.2866j$	$0.0087 + 0.3353j$	$0.0034 - 0.2522j$
	$k_0 + 2$	$-0.0826 - 0.0405j$	$0.0164 - 0.0117j$	$-0.0027 - 0.0042j$
$\{2T, 2T\}$	$k_0 - 2$	$0.0826 + 0.0574j$	$-0.0253 + 0.2920j$	$-0.0826 + 0.0405j$
	$k_0 - 1$	$-0.1241 + 0.2866j$	$-0.6173 - 0.0253j$	$-0.1072 - 0.2866j$
	n_0	$0.4077j$	0.0253	$0.4246j$
	$k_0 + 1$	$0.1241 + 0.2866j$	$-0.6173 + 0.0253j$	$0.1072 - 0.2866j$
	$k_0 + 2$	$-0.0826 + 0.0574j$	$-0.0253 - 0.2920j$	$0.0826 + 0.0405j$

Table 5.2 Coefficients between (m_1, k_1) with adjacent points

	$m_1 - 2$	$m_1 - 1$	m_1	m_0	$m_1 + 2$
$k_1 - 2$	0	0	0	$0.0826 + 0.0574j$	0
$k_1 - 1$	$0.1062j$	$0.2501j$	$0.3183j$	$-0.1241 + 0.2866j$	$0.1062j$
k_1	0	$-0.5004j$	0.9997	$0.4077j$	0
$k_1 + 1$	$-0.1062j$	$0.2501j$	$-0.3183j$	$0.1241 + 0.2866j$	$-0.1062j$
$k_1 + 2$	0	0	0	$-0.0826 + 0.0574j$	0

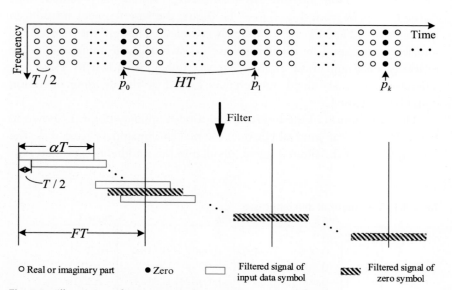

Fig. 5.14 Illustration of segments and subblocks in the S-PTS scheme.

$$
p_d = \begin{cases}
\frac{DT - \alpha T/2}{T/2} + 1 = 2H - \alpha + 1, & d = 0, \\
p_0 + \frac{HT}{T/2}k = 2(1 + d)H - \alpha + 1, & d > 0.
\end{cases}
\tag{5.49}
$$

Compared to the C-PTS scheme, the data rate of the S-PTS scheme is decreased due to the zeros insertion. The ratio of the data rate between the S-PTS and C-PTS schemes is $\frac{2H-1}{2H}$. However, the decreased data rate can be controlled by setting an appropriate length T_s ($T_s = HT$) of each segment.

5.5.3 Simulation Results

To evaluate the performances of the PAPR reduction with the S-PTS method in OQAM/FBMC systems, 10^4 data blocks are randomly

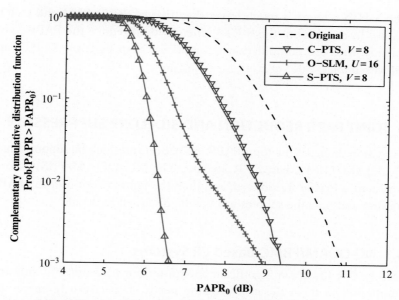

Fig. 5.15 PAPR reductions of the S-PTS, C-PTS, and O-SLM methods.

generated with $K = 64$ and 4QAM, respectively. The rolloff factor of the SRRC filter is 1, and the length of $h(t)$ is $4T$, where $T = 64$.

Fig. 5.15 illustrates the comparison of the PAPR reduction among the S-PTS, C-PTS, and O-SLM schemes. The subblock number for the S-PTS and C-PTS schemes is $V = 8$, and the number of the phase rotation sequences for the O-SLM scheme is $U = 16$. In addition, the segment length of the S-PTS scheme is $T_s = 2T$. It is noted that the PAPR could be reduced by 4.3, 2.0, and 1.1 dB at $CCDF = 10^{-3}$ for the S-PTS, O-SLM, and C-PTS schemes, respectively. Obviously, the PAPR reductions of the S-PTS and O-SLM schemes are better than that of the C-PTS scheme, since the overlapped structure of OQAM/FBMC signals are taken into account in both the S-PTS and O-SLM schemes. Generally, the SLM technique has better PAPR reduction than the PTS technique. However, for the O-SLM scheme, only several previous data blocks are considered, while the following data blocks are ignored. Furthermore, the phases of the first 2α data blocks are unchanged (i.e., the PAPR of the interval $[0, 4T)$ are not reduced). Thus, the PAPR reduction of the S-PTS scheme is better than that of the O-SLM scheme. Moreover, the computational complexity of the S–PTS scheme is much lower than that of the O–SLM and C–PTS schemes. In conclusion, the S–PTS scheme could reduce the PAPR more efficiently with less computational complexity than both the C–PTS and O–SLM schemes.

5.6 JOINT PAPR REDUCTION AND SIDELOBE SUPPRESSION

This section discusses the joint PAPR reduction and sidelobe suppression in NC-OQAM/FBMC-based CR systems. The NC-OQAM/FBMC-based CR system is briefly introduced, and then the joint criterion of the PAPR reduction and sidelobe suppression will be discussed in detail.

5.6.1 NC-OQAM/FBMC-Based CR Systems

Recently, CR has drawn significant attention from academic and industrial communities to meet the ever-growing needs of spectrum resources and high data rate communication [24–26]. For CR systems, noncontiguous filter bank multicarrier (NC-OQAM/FBMC) is an attractive physical layer technology due to its considerable high spectrum efficiency, multipath delay spread tolerance, immunity to the frequency selective fading channels, and high power efficiency. As shown in Fig. 5.16, several PUs and SUs coexist in a typical NC-OQAM/FBMC-based CR wireless system. Here, a subband is available for the SU when the subband is not occupied by the PU. If a subband is available, the NC-OQAM/FBMC-based CR system allows the SU to transmit data under the condition that the interference to the PU is kept at an acceptable level; otherwise, the SU cannot utilize this subband and the wireless system has to reserve some spectrum bands as guard bands to prevent the interference from the SU to the PU. As shown in Fig. 5.16, it is obvious that the NC-OQAM/FBMC technique provides much lower sidelobe power and requires much narrower guard bands than the NC-OFDM technique. As a result, the NC-OQAM/FBMC technique achieves much higher spectrum efficiency and is a much better candidate for CR networks when compared with the NC-OFDM technique. However, the PAPR reduction methods may lead to the increase of the sidelobe power in the NC-OQAM/FBMC-based CR system, resulting in serious interferences from SUs to PUs. Thus, the PAPR and sidelobe power in the NC-OQAM/FBMC-based system need to be jointly reduced.

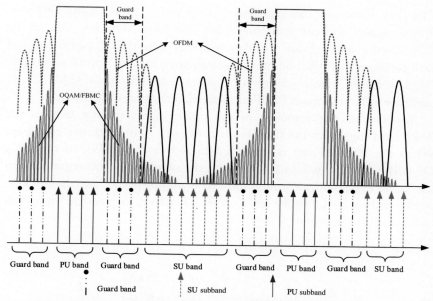

Fig. 5.16 Noncontiguous OQAM/FBMC-based CR wireless system coexisting with the PUs and SUs.

5.6.2 Joint Criterion of PAPR Reduction and Sidelobe Suppression

The criterion of the PAPR reduction and sidelobe suppression in the NC-OQAM/FBMC-based CR system is to search a method which reduces the PAPR significantly, while achieves good sidelobe power performance. Thus, the object of the criterion is to minimize the PAPR, while keeping sidelobe power of the NC-OQAM/FBMC signals at an acceptable level after the operation of the PAPR reduction. Therefore, the mathematical formulation of the criterion is defined as

$$s'(t) = \min \ \mathrm{PAPR}[s(t)],$$
$$\text{subject to: } S \leq \Phi, \tag{5.50}$$

where $s'(t)$ is the time-domain NC-OQAM/FBMC signal with joint PAPR reduction and sidelobe suppression, S is the spectral sidelobe power in the adjacent PU frequency bands, and Φ denotes the allowed spectrum mask.

Some PAPR reduction methods have been investigated for OQAM/FBMC systems, and one of the most attractive PAPR techniques is

PTS [11], because it can significantly reduce the PAPR of the OQAM/
FBMC signals without introducing any distortion. The key idea of the
PTS scheme is to divide the data symbols into several partitions, and
each partition multiplies different phase rotation factors. Several alternative
OQAM/FBMC signals can then be generated, and the OQAM/FBMC
signal with the minimum PAPR is selected as the transmitted signal.

To illustrate the principle of the criterion that jointly reduces the
PAPR and suppresses the sidelobe, here we give some intuitive examples.
When the PTS scheme is employed to reduce the PAPR of the NC-
OQAM/FBMC signals in CR systems, the objective of the criterion is to
search the signal with the minimum PAPR among all AS, and the sidelobe
power of this signal must be less than Φ. Therefore, the specific expression
of Eq. (5.50) is rewritten as

$$q' = \arg \min_{1 \leq q \leq Q} \ \text{PAPR}[s_q(t)],$$

$$\text{subject to: } S_q = \frac{\bar{P}_q}{2K} \int_{f \in \mathcal{P}} \left| \sum_{k=0}^{K-1} S_{hh} \left(f - \frac{2\pi k}{K} \right) \right| df \leq \Phi, \tag{5.51}$$

where q' is index of the chosen signal, $s_q(t)$ is the qth alternative NC-
OQAM/FBMC signal, S_q is the spectral sidelobe power of $s_q(t)$ in the
adjacent PU frequency bands, and $\bar{P}_q$ is the average power of $s_q(t)$. Note
that there are Q AS for the PTS scheme. K denotes the number of the subcarriers in the NC-OQAM/FBMC-based CR system, and S_{hh} denotes the
power spectrum density of the prototype filter. Note that $\mathcal{P}$ is the set of the
adjacent PU spectrum band. It is obvious and reasonable from Eq. (5.51)
that the spectral sidelobe of the transmitted signal is determined by both
the prototype filter and input NC-OQAM/FBMC signals, which is fully
different from the NC-OFDM-based CR systems with the sidelobe power
determined only by the input NC-OQAM/FBMC signals.

For the PTS scheme, each data block is firstly divided into V subblocks.
For example, the mth data block $\mathbf{x}_m = [x_m^0, x_m^1, \ldots, x_m^k, \ldots, x_m^{K-1}]$ is
divided into V subblocks, that is, $\{\mathbf{x}_m^0, \mathbf{x}_m^1, \ldots, \mathbf{x}_m^v, \ldots, \mathbf{x}_m^{V-1}\}$. Then, the
time-domain signal $\tilde{x}_m^v(t)$ of the vth subblock can be obtained according
to Eq. (5.2). Furthermore, multiply the time-domain signal of the vth sub-
block with the phase rotation sequence $\boldsymbol{\gamma}_m = [\gamma_m^0, \gamma_m^1, \ldots, \gamma_m^v, \ldots, \gamma_m^{V-1}]$,
to generate the AS of the mth data block. In addition, the phase rotation
factor $\gamma_m^v \in \{-1, 1\}$. Finally, one can obtain $Q = 2^V$ alternative time-
domain NC-OQAM/FBMC signals by adding all time-domain signals of
different data blocks.

Algorithm 5.1   Joint PAPR Reduction and Sidelobe Suppression

Initialize V;

Divide the data block x_m into V subblocks to obtain x_m^v and $\tilde{x}_m^v(t)$;

Obtain the qth alternative NC-OQAM/FBMC signal $s_q(t)$;

Compute S_q according to Eq. (5.51), and choose the proper prototype filter. Then, save all the signals that satisfy $S_q \le \Phi$ and omit the other signals;

Select the signal with the minimum PAPR as the transmitted signal among the remaining NC-OQAM/FBMC signals.

For the qth alternative NC-OQAM/FBMC signal $s_q(t)$, one can firstly calculate its sidelobe power S_q, and choose the proper prototype filter. Then, among all the alternative NC-OQAM/FBMC signals, keep all the signals that satisfy $S_q \le \Phi$ and omit the other signals. Moreover, among all the remaining NC-OQAM/FBMC signals, the signal with the minimum PAPR is selected as transmitted signal. In this way, the PAPR of the NC-OQAM/FBMC-based CR system can be significantly reduced, while the good sidelobe suppression performance can be maintained.

Therefore, by employing PTS for NC-OQAM/FBMC-based CR systems, the joint PAPR reduction and sidelobe suppression approach can be summarized in Algorithm 5.1.

5.6.3 Simulation Results

This subsection conducts simulations to quantify the capability of NC-OQAM/FBMC-based CR system with the criterion, including performances of the PAPR minimization and sidelobe suppression. An NC-OQAM/FBMC-based CR system is considered, where the number of subcarriers is $K = 64$, among which 54 subcarriers are supposed to be utilized by the SUs (from the 0th subcarrier to the 29th subcarrier, and from the 40th subcarrier to the 63th subcarrier) to transmit data, and the other 10 subcarriers are occupied by the PU. Moreover, the target PU spectrum band is from $30\Delta f$ to $39\Delta f$, where Δf denotes the subcarrier interval. Furthermore, in the simulations, the NC-OQAM/FBMC-based CR system employs $M = 100$ data blocks which are modulated by 4-QAM modulation. Note that the spectrum density of the NC-OQAM/FBMC-based CR system is calculated using the Welch method with Blackman window.

For the criterion, the sidelobe power is constrained to be less than Φ, and the sidelobe power is determined by both the signal power and the prototype filter. Thus, one can choose the proper prototype filter to constrain the sidelobe power to be less than Φ. As a result, the criterion achieves good PAPR performance while constraining the sidelobe power to be under an acceptable level. In the simulations, the criterion employs the PTS method with $V = 4$ and $Q = 16$ to reduce the PAPR.

Fig. 5.17 shows the PAPR reduction of the NC-OQAM/FBMC-based CR system with the criterion. It is obvious that the NC-OQAM/FBMC-based CR system with the criterion offers much better PAPR reduction performance than the original NC-OQAM/FBMC-based CR system without PAPR reduction. Compared with the original NC-OQAM/FBMC-based CR system, the PAPR reduction scheme with the criterion can reduce the PAPR by 2.8 dB when $CCDF = 10^{-2}$.

Fig. 5.18 investigates the sidelobe suppression performance of the criterion with $\Phi = -60$ dB. For the original NC-OQAM/FBMC-based CR system, the raised-cosine filter [27] (denoted as $h_1(t)$) is employed. However, the original NC-OQAM/FBMC-based CR system

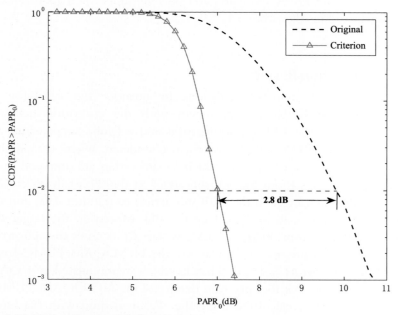

Fig. 5.17 PAPR reduction with the criterion.

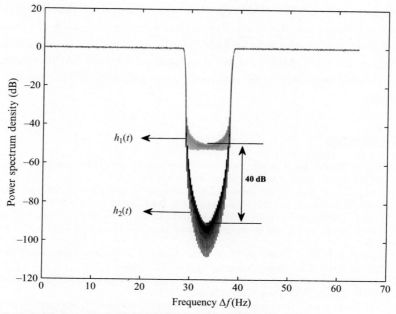

Fig. 5.18 Sidelobe suppression with the criterion with $\Phi = -60$ dB.

with raised–cosine filter cannot provide very good sidelobe performance. For the criterion, the proper prototype filter is adopted to fulfill the sidelobe power constraint in Eq. (5.51). When $\Phi = -60$ dB, the raised–cosine filter cannot satisfy the sidelobe power constraint, while the sidelobe power constraint can be satisfied if the optimized filter (denoted as $h_2(t)$) in [28] is employed. As shown in Fig. 5.18, the original NC-OQAM/FBMC system adopting $h_1(t)$ can achieve the spectrum notch of about -50 dB in the PU spectrum band, and it cannot satisfy the sidelobe power constraint. Moreover, the criterion adopting $h_2(t)$ can achieve the spectrum notch of about -90 dB in the PU spectrum band, thus the sidelobe power constraint can be satisfied. In this way, the criterion can improve the PAPR reduction while providing satisfied sidelobe performance.

5.7 SUMMARY

In this chapter, we placed an emphasis on summarizing most of the existing PAPR reduction methods for OQAM/FBMC systems. The high PAPR in OQAM/FBMC systems was firstly introduced. Then, the PAPR distribution of OQAM/FBMC signals was derived and the PAPR effect on power amplifiers was discussed. Furthermore, we investigated all efficient PAPR reduction methods in OQAM/FBMC systems (i.e., the AS, MBJO, and segmental PTS). Moreover, we also discussed the joint PAPR reduction and sidelobe suppression for noncontiguous OQAM/FBMC-based CR networks.

REFERENCES

[1] Jiang T, Wu Y. An overview: peak-to-average power ratio reduction techniques for OFDM signals. IEEE Trans Broadcast 2008;54(2):257–68.

[2] Li H, Jiang T, Zhou Y. An improved tone reservation scheme with fast convergence for PAPR reduction in OFDM systems. IEEE Trans Broadcast 2011;57(4):902–6.

[3] Jiang T. Orthogonal frequency division multiplexing technology in communication systems. In: Encyclopedia of wireless and mobile communications, USA: CRC Press; 2007.

[4] Gutman I, Iofedov I, Wulich D. Iterative decoding of iterative clipped and filtered OFDM signal. IEEE Trans Commun 2013;61(10):4284–93.

[5] Wang YC, Luo ZQ. Optimized iterative clipping and filtering for PAPR reduction of OFDM signals. IEEE Trans Commun 2011;59(1):33–7.

[6] Li L, Qu D, Jiang T. Partition optimization in LDPC-coded OFDM systems with PTS PAPR reduction. IEEE Trans Veh Technol 2014;63(8):4108–13.

[7] Qu D, Li L, Jiang T. Invertible subset LDPC code for PAPR reduction in OFDM systems with low complexity. IEEE Trans Wirel Commun 2014;13(4):2204–13.

[8] Jiang T, Zhu G. Nonlinear companding transform for reducing peak-to-average power ratio of OFDM signals. IEEE Trans Broadcast 2004;50(3):342–6.

[9] Jiang T, Yang Y, Son Y. Exponential companding technique for PAPR reduction in OFDM systems. IEEE Trans Broadcast 2005;51(2):244–8.

[10] Jiang T, Yao W, Guo P, Song Y, Qu D. Two novel nonlinear companding schemes with iterative receiver to reduce PAPR in multi-carrier modulation systems. IEEE Trans Broadcast 2006;51(2):268–73.

[11] Li L, Qu D. Joint decoding of LDPC code and phase factors for OFDM systems with PTS PAPR reduction. Trans Veh Technol 2013;62(1):444–9.

[12] Jiang T, Li C. Simple alternative multi-sequences for PAPR reduction without side information in SFBC MIMO OFDM systems. Trans Veh Technol 2012;61(7):3311–5.

[13] Goff S, Khoo B, Tsimenidis C, Sharif B. A novel selected mapping technique for PAPR reduction in OFDM systems. IEEE Trans Commun 2008;56(11):1775–9.

[14] Krongold B, Jones D. PAR reduction in OFDM via active constellation extension. IEEE Trans Broadcast 2003;49(3):258–68.

[15] Wang Y, Chen W, Tellambura C. Genetic algorithm based nearly optimal peak reduction tone set selection for adaptive amplitude clipping PAPR reduction. IEEE Trans Broadcast 2012;58(3):462–71.

[16] Damavandi M, Abbasfar A, Michelson D. Peak power reduction of OFDM systems through tone injection via parametric minimum cross-entropy method. IEEE Trans Veh Technol 2013;62(4):1838–43.

[17] Zhou Y, Jiang T, Huang C, Cui S. Peak-to-average power ratio reduction for OFDM/OQAM signals via novel alternative signals method. IEEE Trans Veh Technol 2014;63(1):494–9.

[18] Qu D, Lu S, Jiang T. Multi-block joint optimization for the peak-to-average power ratio reduction of FBMC-OQAM signals. IEEE Trans Signal Process 2013;61(7): 1605–17.

[19] Ye C, Li Z, Jiang T, Ni C, Qi Q. PAPR reduction of OQAM-OFDM signals using segmental PTS scheme with low complexity. IEEE Trans Broadcast 2014;60(1):141–7.

[20] Jiang T, Ni C, Qu D, Wang C. Energy-efficient NC-OFDM/OQAM-based cognitive radio networks. IEEE Commun Mag 2014;52(7):54–60.

[21] Cann AJ. Nonlinearity model with variable knee sharpness. IEEE Trans Aerosp Electron Syst 1980;16(6):874–7.

[22] Skrzypczak A, Javaudin J, Sinhan P. Reduction of the peak-to-average power ratio for the OFDM/OQAM modulation. In: IEEE VTC, Melbourne, Australia; 2006.

[23] Zhou GT, Peng L. Optimality condition for selected mapping in OFDM. IEEE Trans Signal Process 2006;54(8):3159–65.

[24] Haykin S. Cognitive radio: brain-empowered wireless communications. IEEE J Sel Areas Commun 2005;23(2):201–20.

[25] Zou Y, Yao Y, Zheng B. Cooperative relay techniques for cognitive radio systems: spectrum sensing and secondary user transmissions. IEEE Commun Mag 2012;50(4): 98–103.

[26] Qu D, Wang Z, Jiang T. Extended active interference cancellation for sidelobe suppression in cognitive radio OFDM systems with cyclic prefix. IEEE Trans Veh Technol 2010;59(4):1689–95.

[27] Alagha NS, Kabal P. Generalized raised-cosine filters. IEEE Trans Commun 1999; 47(7):989–97.

[28] Chen D, Qu D, Jiang T. Prototype filter optimization to minimize stopband energy with NPR constraint for filter bank multicarrier modulation systems. IEEE Trans Signal Process 2013;61(1):159–69.

CHAPTER 6

Overhead Reduction

Offset quadrature amplitude modulation-based filter bank multicarrier (OQAM/FBMC) [1–7] is considered as a promising alternative to the conventional orthogonal frequency division multiplexing (OFDM) technique [8, 9]. However, in burst transmission mode, the filters of OQAM/FBMC systems introduce transitions at the beginning and end of the burst, which extends the length of the burst and incurs certain penalty in spectral efficiency.

Several techniques for tail shortening in OQAM/FBMC systems have already been put forward in literatures. One traditional method to resolve this problem is hard truncation of the tail [10]. Although it totally removes the tail, the truncation causes intersymbol interference/intercarrier interference (ISI/ICI) on the data symbols, since it distorts the symbols close to the edge of the burst. Moreover, the method increases the out-of-band (OOB) emission due to sharp truncation of signal. An improved method uses a generalized weighting window to smoothen the edge transitions introduced by truncation, which improves the OOB performance, however, it causes more ISI/ICI than the hard truncation method [11]. In [12], a different approach called weighted circular convolution was proposed, which circularly extends the finite-length OQAM/FBMC sequence to infinite-length and applies a weighted time-windowing to resolve the edge transitions of the overhead-removed signal. Although the approach does not incur any ISI/ICI, it requires that the channel remains unchanged over the time of a frame, that is, the frame period is shorter than the channel correlation time. Another method aims to use a package transmission concept with special processing at the edge side [12]. Although this method does not need truncation, which avoids the performance loss, it is only valid for short filter lengths.

This chapter presents a novel method based on virtual symbols for tail shortening of OQAM/FBMC signals. Virtual symbols refer to the symbols that do not convey any data and are assumed to be zero at the receiver.

OQAM/FBMC for Future Wireless Communications
http://dx.doi.org/10.1016/B978-0-12-813557-0.00006-1

© 2018 Elsevier Ltd.
All rights reserved.

The tail-shortening method transmits certain signals on the virtual symbols to cancel most of the tail, which results in no interference on the data symbols. This chapter uses two approaches, that is, hard truncation and truncation with windowing, to deal with the residual tail. This chapter then presents an optimization method that allows computation of virtual symbols for each data packet.

The rest of this chapter is organized as follows. Section 6.1 describes tail truncation methods. The tail-shortening method by virtual symbols is presented in Section 6.2. Section 6.3 describes the optimization method. Finally, a summary is given in Section 6.4.

6.1 TAIL TRUNCATION METHODS

Since the OQAM/FBMC systems adopt filters with rather long impulse responses, the resulting ramps at the edges of each data burst cover multiple symbol intervals, which considerably reduces the gains of OQAM/FBMC in spectral efficiency.

In what follows, this chapter uses the tail at the end of the burst to illustrate the tail-shortening methods. The processing at the start of the burst is similar.

Assume that the discrete-time OQAM/FBMC signal of the transmitter is a burst, which is written as

$$s(n) = \sum_{k \in \Omega} \sum_{m=0}^{N_f-1} d_k(m) h\left(n - m\frac{K}{2}\right) e^{j2\pi kn/K} e^{j(k+m)\pi/2}, \qquad (6.1)$$

where Ω is the set of N_Ω active subcarriers, N_f is the number of real-valued symbols, and $h(n)$ is the impulse response of the prototype filter. The length of the prototype filter (i.e., $L_p = \alpha K + 1$) depends on the size of the subcarriers K and the overlapping factor α. The burst length L_b is then $(N_f - 1)T/2 + \alpha T$, where T is the complex-valued symbol interval.

One traditional method to resolve the problem is hard truncation of the tail, which obtains a tail-removed signal as

$$s_{h.trunc}(n) = \begin{cases} s(n), & n \le K_e \\ 0, & n > K_e \end{cases}, \qquad (6.2)$$

where K_e is the point of truncation. The cost of this method is increased OOB radiation and ISI/ICI due to the truncation.

Another method called truncation with windowing uses a generalized weighting window to remove the tail. The windowed signal is given as

$$s_{\text{w.trunc}}(n) = s(n)w(n),\tag{6.3}$$

where $w(n)$ is the window function. Specifically, this chapter considers a raised–cosine window, which is expressed as

$$w(n) = \begin{cases} 1, & n < K_{\text{b.ro}}, \\ \frac{1}{2} + \frac{1}{2}\cos\left(\frac{(n-K_{\text{b.ro}})\pi}{L_r}\right), & K_{\text{b.ro}} \leq n \leq K_e, \\ 0, & n > K_e, \end{cases}\tag{6.4}$$

where L_r is the length of roll-off and $K_{\text{b.ro}} = K_e - L_r$ is the beginning of the roll-off. Although the method smoothens the edge transitions introduced by truncation, it causes greater ISI/ICI than the hard truncation when both methods truncate the signal at K_e.

A variation of truncation with windowing is the so-called circular convolution with windowing [12], which makes the OQAM/FBMC signal periodical and truncates it with windowing. This method has very low ISI/ICI due to the fact that the original OQAM/FBMC signal can be recovered from a single period of the circular convolved signal and the signal affected by window does not participate in the detection at the receiver. However, the circular convolution makes the burst of OQAM/FBMC signal a single structured block, which is more sensitive to carrier frequency offset and Doppler drift, and also complicates the reception.

6.2 TAIL SHORTENING BY VIRTUAL SYMBOLS

The basic idea is to transmit certain signals on the virtual symbols to cancel most of the tail. Virtual symbols refer to the symbols that are beyond the scope of data symbols for a burst, as demonstrated in Fig. 6.1. Virtual symbols do not convey any data and are assumed to be zero at the receiver. They are also real-valued symbols as the data symbols, and the cancelation signal is generated by modulating them as pulse amplitude modulated (PAM) symbols onto the corresponding subcarriers.

In this section, we present the tail-shorterning method in three major steps.

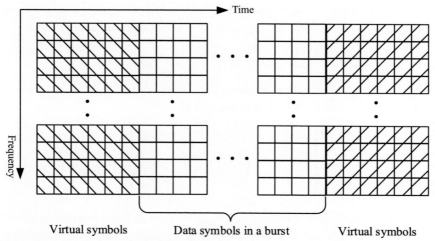

Fig. 6.1 Virtual symbols in an OQAM/FBMC burst.

6.2.1 Tail Analysis

Firstly, define the tail $s_t(n)$ as

$$s_t(n) = \begin{cases} 0, & n < K_t \\ s(n), & n \geq K_t \end{cases}, \tag{6.5}$$

where K_t is the time index of the start of the tail. The tail $s_t(n)$ is then analyzed by the filter bank and the result is matrix $\mathbf{V}_\Omega$, whose (p, q)th element, $v(p, q)$, is given by

$$v(p, q) = \begin{cases} \Re\left\{\sum_{n=-\infty}^{\infty} s_t(n)h(n - q\frac{K}{2})e^{\frac{-j2\pi pn}{K}}e^{\frac{-j(p+q)\pi}{2}}\right\}, & p \in \Omega, N_f \leq q \leq N_f + N_V - 1 \\ 0, & \text{else,} \end{cases} \tag{6.6}$$

where $\Re(\cdot)$ represents real part, and N_V is a positive integer number. Rows and columns of matrix $\mathbf{V}_\Omega$ represent frequency and time, respectively. $v(p, q)$'s with q from time index N_f to $N_f + N_V - 1$ are referred to as *virtual symbols* throughout this chapter, for the reason that they do not convey any data and are assumed to be zero at the receiver. In the tail analysis, only the virtual symbols have nonzero outputs, while outputs of other symbols are forced to be zero.

6.2.2 Tail Reconstruction With Virtual Symbols

This section uses the matrix $\mathbf{V}_\Omega$ to form the cancelation signal $s_{vs}(n)$ as the reconstructed tail, that is,

$$s_{vs}(n) = \sum_{p\in\Omega} \sum_{q=N_f}^{N_f+N_V-1} v(p,q)h\left(n - q\frac{K}{2}\right) e^{j2\pi pn/K} e^{j(p+q)\pi/2}. \quad (6.7)$$

After applying $s_{vs}(n)$ to the original OQAM/FBMC signal, the tail-canceled signal can be obtained as

$$s_{vs.short}(n) = s(n) + s_{vs}(n). \quad (6.8)$$

Since the cancelation signal $s_{vs}(n)$ is constructed following the same way as the original OQAM/FBMC signal burst, it has no interference on the data symbols.

To demonstrate the effect of the distortionless tail cancelation, this section simulates a burst of $N_f = 40$ real-valued symbols (20 complex OQAM symbols) and $N_V = 6$ virtual symbols. The number of the subcarriers is set as $K = 256$ and the number of active subcarriers is set as $N_\Omega = 200$. This chapter uses the PHYDYAS [13–15] filter with length $L_p = 4K + 1$. We set $K_t = K_{f.symbol} + T/4$, where $K_{f.symbol}$ is the time corresponding to the center of the final real-valued symbol (the $(N_f - 1)$th real-valued symbol). Fig. 6.2 shows the magnitude of the original OQAM/FBMC signal $s(n)$ and the tail-canceled signal $s_{vs.short}(n)$. It is observed that most of the tail after K_t has been significantly suppressed.

6.2.3 Truncation With Windowing

Although the tail of the signal $s_{vs.short}(n)$ is significantly suppressed, it is not zero and needed to be truncated before sending to the output. By applying a generalized weighting window to remove the residual tail part, the output signal is obtained as

$$s_{vs.trunc}(n) = s_{vs.short}(n)w(n), \quad (6.9)$$

where $w(n)$ is the window function.

The window function $w(n)$ may be a raised-cosine window as in Eq. (6.4), or one may choose to hard truncate $s_{vs.short}$ at K_e. When the raised-cosine window is considered, let $L_r = K_e - K_t$ and choose K_t such that the roll-off length L_r be the same as its counterpart in the conventional windowing method. This will allow a fair comparison.

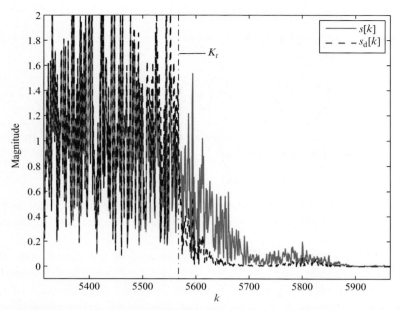

Fig. 6.2 Magnitude of the original OQAM/FBMC signal $s(n)$ and the tail-canceled signal $s_{\text{vs.short}}(n)$.

6.2.4 Performance Evaluation

In this section, the performance of the tail-shortening method is evaluated via simulations. The simulation parameters are the same as those of Section 6.2.2, except for those stated in the following. Simulation results of hard truncation and truncation with windowing methods are also presented for comparison. A raised-cosine window with roll-off length L_r is applied to the truncation with windowing method.

To define the tail overhead of OQAM/FBMC signals, we consider a reference OFDM signal with the same number of complex symbols $(N_f/2)$ and without CP. The signal has a length of $N_f T/2$, or it ends at $K_{\text{ref}} = K_{\text{f.symbol}} + T/4$. Next, define the tail overhead of OQAM/FBMC signal with respect to K_{ref} as: $L_{\text{oh}} = K_e - K_{\text{ref}}$, where K_e is taken as the tail end of the OQAM/FBMC signal with tail shortenings.

Fig. 6.3 shows the power spectral density (PSD) of the tail-shortened signals for $L_{\text{oh}} = L_r = T/4$. As observed, in terms of OOB emission, the tail shortening with virtual symbols and windowing methods have the same performances when both of them use the same raised-cosine window. The tail shortening with virtual symbols and hard limiting also has a comparable

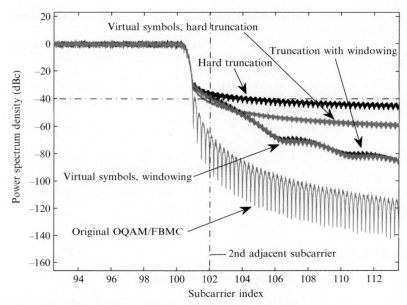

Fig. 6.3 Comparison of PSD with $L_{oh} = T/4$ and $L_r = T/4$.

performance. It has a better OOB emission at the beginning of the stop band, but degrades at the later part.

Besides the OOB emission, another measure of qualifying the various tail-shortening methods is to quantify them by looking at the introduced distortion in the recovered data symbols. This distortion is quantified by the error vector magnitude (EVM) of data symbols at the beginning and the end (i.e., the edge symbols) of each burst. Table 6.1 lists EVM values of the edge symbols for different tail-shortening methods, $L_r = T/4$, and three choices of $L_{oh} = T/4$, $3T/8$, and $T/2$. As seen, the tail shortening with virtual symbols outperforms both methods of hard truncation and windowing. It is also interesting to note that, with respect to EVM, hard truncation outperforms truncation with windowing. This is because, for a set value of K_e, windowing, in addition to hard truncation, adds some distortion to the first and last L_r samples of each burst.

From the earlier simulation results, the effects of the tail cancelation by virtual symbols are clearly observed. First, most of the tail signal is canceled without generating any interference to the data symbols, therefore the distortion resulted from latter windowing is minimized. Second, the distortionless tail cancelation signal is a spectrally well-contained in-band

Table 6.1 EVM of the edge symbols for different tail-shortening methods, and the choice of $L_r = T/4$

L_{oh}	Hard	Windowing	Virtual symbols with windowing	Virtual symbols with hard truncation
$T/4$	$-29.59\,\text{dB}$	$-22.48\,\text{dB}$	$-35.02\,\text{dB}$	$-48.16\,\text{dB}$
$3T/8$	$-41.03\,\text{dB}$	$-30.46\,\text{dB}$	$-48.19\,\text{dB}$	$-60.99\,\text{dB}$
$T/2$	$-49.01\,\text{dB}$	$-41.20\,\text{dB}$	$-62.46\,\text{dB}$	$-66.78\,\text{dB}$

signal; therefore, it has very little influence on the OOB emission. Due to the earlier reasons, it is not surprising to find that the method has a much improved EVM and similar OOB emission performance, compared with the windowing method with the same window.

6.3 OPTIMIZATION METHOD

6.3.1 Tail Cancelation by Virtual Symbols

To find the virtual symbols, that is, the elements of $\mathbf{V}_\Omega$ define the cost functions

$$\xi_1 = \sum_{K_e < n \leq K_e + L_1} |s_{vs.short}(n)|^2, \tag{6.10}$$

and

$$\xi_2 = \sum_{K_e - L_2 + 1 \leq n \leq K_e} |s_{vs}(n)|^2, \tag{6.11}$$

and search for a proper $\mathbf{V}_\Omega$ that minimizes ξ_1 and ξ_2 jointly, where L_2 and L_1 are the time domain duration of the virtual symbols before and after the point K_e, respectively. Minimization of ξ_1 suppresses the tail of the burst, beyond K_e. Minimization of ξ_2, on the other hand, is to assure that the total energy of $s_{vs}(n)$ over the range of $n \leq K_e$ is kept minimal. This is necessary to avoid an unexpected increase of $|s_{vs.short}(n)|$ over this range, hence, avoid an undesirable increase of PAPR of the burst.

To obtain a balanced minimization of ξ_1 and ξ_2, we find $\mathbf{V}_\Omega$ that minimizes the combined cost function

$$\xi = \xi_1 + \gamma \xi_2, \tag{6.12}$$

where γ is a positive parameter that should be found empirically. To derive a convenient formulation for this minimization problem, we proceed as follows.

We define the pair of column vectors $\mathbf{s}_1$ and $\mathbf{s}_2$ consisting of elements $s_{\text{vs.short}}(n)$, for $K_e < n \leq K_e + L_1$ and $s_{\text{vs}}(n)$, for $K_e - L_2 + 1 \leq n \leq K_e$, respectively. We also note that $\mathbf{s}_1$ and $\mathbf{s}_2$ can be expanded as

$$\mathbf{s}_1 = \mathbf{s} + \mathbf{W}_\alpha \mathbf{v}_\Omega, \tag{6.13}$$

and

$$\mathbf{s}_2 = \mathbf{W}_\beta \mathbf{v}_\Omega, \tag{6.14}$$

where $\mathbf{s}$ contains samples of $s(n)$, for $n > K_e$, $\mathbf{v}_\Omega$ is constructed by rearranging the columns of $\mathbf{V}_\Omega$ in a column vector, and $\mathbf{W}_\alpha$ and $\mathbf{W}_\beta$ are L_1-by-$N_\Omega N_V$ and L_2-by-$N_\Omega N_V$ matrices, respectively, which could be obtained trivially by taking note of the relationship in Eq. (6.7). Similarly, $\mathbf{s} = \mathbf{W}_0 \mathbf{a}$, where $\mathbf{a}$ is a vector consisting of $d_k(m)$ for $k \in \Omega$ and $N_f - N_P \leq m \leq N_f - 1$ and $\mathbf{W}_0$ is an L_1-by-$N_\Omega N_P$ matrix. Here, we assume that $\mathbf{a}$ consists of the PAM symbols that contribute to the tail.

Substituting Eqs. (6.13), (6.14) into Eqs. (6.10), (6.11), respectively, and considering Eq. (6.12), we can obtain

$$\xi = \|\mathbf{s} + \mathbf{W}_\alpha \mathbf{v}_\Omega\|^2 + \gamma \|\mathbf{W}_\beta \mathbf{v}_\Omega\|^2, \tag{6.15}$$

Next, noting that $\mathbf{a}$ is real-valued, but $\mathbf{s}$, $\mathbf{W}_\alpha$, and $\mathbf{W}_\beta$ are complex-valued, one may rearrange Eq. (6.15) as

$$\xi = \left\| \begin{bmatrix} \tilde{\mathbf{s}} \\ \mathbf{0} \end{bmatrix} + \begin{bmatrix} \tilde{\mathbf{W}}_\alpha \\ \sqrt{\gamma}\tilde{\mathbf{W}}_\beta \end{bmatrix} \mathbf{v}_\Omega \right\|^2, \tag{6.16}$$

where

$$\tilde{\mathbf{s}} = \begin{bmatrix} \Re[\mathbf{s}] \\ \Im[\mathbf{s}] \end{bmatrix}, \quad \tilde{\mathbf{W}}_\alpha = \begin{bmatrix} \Re[\mathbf{W}_\alpha] \\ \Im[\mathbf{W}_\alpha] \end{bmatrix}, \quad \tilde{\mathbf{W}}_\beta = \begin{bmatrix} \Re[\mathbf{W}_\beta] \\ \Im[\mathbf{W}_\beta] \end{bmatrix},$$

$\Re[\cdot]$ and $\Im[\cdot]$ denote the real and imaginary parts, respectively, and $\mathbf{0}$ indicates a zero column vector.

The minimization of ξ is a least squares problem, whose solution can be expressed as

$$\mathbf{v}_\Omega = -\mathbf{B}\tilde{\mathbf{s}}, \tag{6.17}$$

where

$$\mathbf{B} = \left(\tilde{\mathbf{W}}_\alpha^T \tilde{\mathbf{W}}_\alpha + \gamma \tilde{\mathbf{W}}_\beta^T \tilde{\mathbf{W}}_\beta \right)^{-1} \tilde{\mathbf{W}}_\alpha^T. \tag{6.18}$$

Using $\mathbf{s} = \mathbf{W}_0\mathbf{a}$, where

$$\tilde{\mathbf{W}}_0 = \left[\begin{array}{c} \Re[\mathbf{W}_0] \\ \Im[\mathbf{W}_0] \end{array} \right],$$

we have

$$\mathbf{v}_\Omega = -\mathbf{B}\tilde{\mathbf{W}}_0\mathbf{a}. \tag{6.19}$$

We also note that $-\mathbf{B}\tilde{\mathbf{W}}_0$ is a fixed matrix that depends on the prototype filter $h(n)$ and parameters N_Ω, N_P, N_V, and γ. The matrix $-\mathbf{B}\tilde{\mathbf{W}}_0$ thus can be precalculated and used in Eq. (6.19) to obtain $\mathbf{v}_\Omega$ for each data packet.

6.3.2 Truncation of Residual Tail

Energy of $s_{vs.short}(n)$ beyond K_e will be very small, after canceling the tail, if K_e and γ are selected properly. However, the tail will not be exactly zero due to limited number of virtual symbols and constraint of ξ_2. Regarding the residual tail, either one of the following two approaches could be applied: (1) Truncate the residual tail beyond K_e. This could raise the OOB spectrum and incur interference to the edge data symbols. (2) Keep the residual tail and allow overlapping of the successive packets. Such overlap, clearly, leads to interpacket interference. However, our numerical study reveals that such interference, even when the design parameters are selected correctly, remains negligible.

Regarding the truncation, two approaches (i.e., hard truncation and truncation with windowing) are available. For hard truncation, we simply throw away the residual tail beyond K_e. While for the truncation with windowing, the cost functions in the previous section are slightly modified as the following:

$$\xi_1 = \sum_{K_b < n \le K_e + L_1} |[1 - w(n)]s_{vs.short}(n)|^2, \tag{6.20}$$

and

$$\xi_2 = \sum_{K_e - L_2 + 1 \le n \le K_e} |w(n)s_{vs}(n)|^2, \tag{6.21}$$

where the roll-off of window $w(n)$ is defined from $K_{b.ro}$ to K_e. These modification reflects the additional distortions due to the windowing.

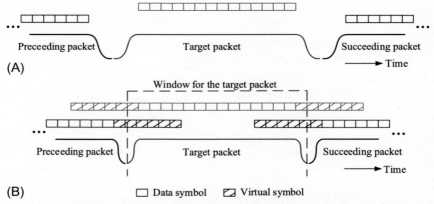

Fig. 6.4 Overlapping virtual symbols with the adjacent bursts. (A) Without tail shortening. (B) With tail shortening by virtual symbols.

6.3.3 Overhead and Complexity

In this section, we discuss the overhead and complexity issues of the optimization method.

Overhead

It may look like that adopting the virtual symbols would decrease the spectrum efficiency of OQAM/FBMC systems, since the virtual symbols consume additional time-frequency resources. Contrary to this observation, it actually increases the spectrum efficiency by placing the OQAM/FBMC bursts/packets closer to each other as illustrated in Fig. 6.4. Fig. 6.4 shows a sequence of OQAM/FBMC packets, which we assume are from different users and asynchronous in timing. With the tail shortening, successive packets are allowed to be placed much closer in time than those without tail shortening, thus more data is allowed in a give channel. It is clear in this situation that the virtual symbols are overlapped with the virtual symbols and/or data symbols of the adjacent packets, thus they do not consume any additional time-frequency resources and are not overheads as they looks like.

Due to the overlapped virtual symbols, receiver with tail shortening has to be carefully designed to avoid interference from adjacent packets. Such a receiver could work as shown by Fig. 6.4. First, the receiver detects the beginning and end of the target packet, say, K_b and K_e. This could be accomplished by timing synchronization mechanism of the OQAM/FBMC system, which relies on the pilot/preamble symbols.

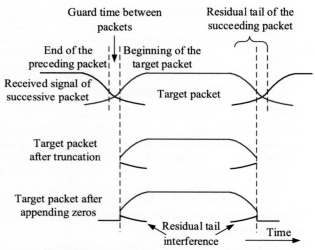

Fig. 6.5 Receiver design with overlapping of residual tails and adjacent packets (that may originate from different transmitters).

Second, the receiver applies a rectangular window to extract the target packet from the received signal (truncating the signal at the beginning and end). Third, the receiver appends certain length of zeros at both sides of the truncated packet and feeds it to a standard OQAM/FBMC receiver.

As discussed at the end of the last section, two approaches could be applied to the residual tail. If the transmitter truncates the residual tail beyond K_b or K_e, there will be no interference from adjacent packets, as long as the guard time between packets is long enough to absorb the channel spread. If the residual tail is not truncated and, to keep the same spectrum efficiency as the truncated case, overlapped with the adjacent packets as shown in Fig. 6.5, it is an interference to the packet it overlaps with. Apparently, this interference is related to the total energy of the residual tail. We will see how the error performance is affected by the residual tail in Section 6.3.4.

Complexity

Complexity of the optimization method comes from two aspects: (1) Calculation of the virtual symbols by multiplying $-\mathbf{B}\tilde{\mathbf{W}}_0$ with $\mathbf{a}$, which has a complexity of $N_P N_V N_\Omega^2$ real multiplications; (2) Transforming virtual symbols into the cancelation signal, whose complexity is proportional to $N_V N_\Omega$. Since the cancelation signal is an OQAM/FBMC signal, the actual complexity of transforming depends on the OQAM/FBMC implementation. It is a fraction of the complexity generating the data signal since the virtual symbols is usually much less than the data symbols.

As discussed earlier, a small N_V is critical for the optimization method to keep an acceptable complexity. Therefore, We will simulate with very small N_V, such as one or two, to see if the method could provide satisfactory error and PSD performance in Section 6.3.4. With N_V as small as one, a good tail cancelation cannot be achieved. In such case, combining the tail shortening and truncation with window provides a not-bad tradeoff between complexity and performance of error and PSD. This will also be evaluated in Section 6.3.4.

Another effective way to reduce the complexity is to keep N_Ω small, since the complexity of calculating the virtual symbols is proportional to N_Ω^2. For an OQAM/FBMC signal with large number of active subcarriers, one may divide the subcarriers into smaller groups and conduct the tail shortening separately within every group. For example, we may divide the 192 subcarriers of an OQAM/FBMC signal into 16 groups with 12 subcarriers each, and conduct the tail-shortening algorithm separately within each group. In this way, the complexity is reduced by a factor of 16, at the cost of slightly degraded error and PSD performance.

6.3.4 Performance Evaluation

In this section, the performance of the optimization method is evaluated via simulations. We simulate an OQAM/FBMC system with a maximum number of subcarriers of $K = 256$ where only a subset of which will be active. The number of active subcarriers is 12 unless otherwise noted.

We employ the PHYDYAS filter and IOTA filter with length $L_p = 4K + 1$. End of the packet is $K_e = K_{f.symbol} + K/2$, where $K_{f.symbol}$ is the time corresponding to the center of the final real-valued symbol (the $(N_f - 1)$th real-valued symbol). The number of virtual symbols (N_V) on each end of the burst is set to 1, 2, or 6 for the optimization method. A raised-cosine window with roll-off length L_r is applied for truncation with windowing, where $L_r = K/4$. In the simulations, K_e is set to $K_{f.symbol} + K/2$, which corresponds to an overhead $L_{oh} = K/4$ or $T/4$ in time on each end.

Each OQAM/FBMC burst consists of $N_f = 14$ real-valued PAM OQAM/FBMC symbols (7 complex OQAM/FBMC symbols), where the tails on both ends add up to an overhead of $100 \times (2 \times T/4)/(7 \times T) = 7.14\%$ of the duration of the seven complex symbols. We selected the burst size to match the smallest resource unit in the LTE-OFDM standard [16]. For this choice, the OQAM/FBMC overhead closely matches the CP overhead of LTE-OFDM which is 7.03% when a minimum length CP (often referred to as normal CP) is applied. We note that in OFDM this overhead increases to 25% if an extended CP [16] is used. On the other hand, in OQAM/FBMC the above overhead decreases in longer data packets.

Residual Tail and Cancelation Signal Energy

Fig. 6.6 shows the average residual tail and cancelation signal energy (ξ_1 and ξ_2, respectively) for a range of γ. Here, we have used PHYDAYS filter and the number of active subcarriers is equal to 12. The residual tail and cancelation signal energy are normalized with respect to the energy of each real-valued PAM OQAM/FBMC symbol in the OQAM/FBMC burst (the dBc unit in Fig. 6.6 refers to decibels relative to the energy of each real-valued PAM OQAM/FBMC symbol).

As observed, $\gamma = 0.1$ provides a good balance between the residual tail and cancelation signal energy for all N_V's considered. At this γ, tail cancelation with $N_V = 2$ already provides a close performance to that with a very large N_V, such as 6. For instance, at $\gamma = 0.1$ the residual tail

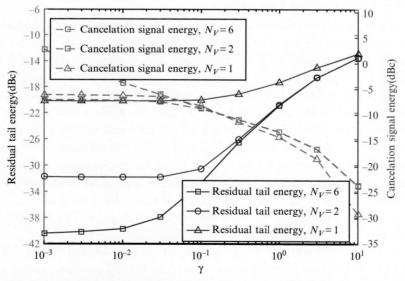

Fig. 6.6 Average residual tail and cancelation signal energy, normalized by the energy of one real-valued PAM OQAM/FBMC symbol in the OQAM/FBMC burst.

energy drops below -30 dBc, and the cancelation signal energy is below -20 dBc, for $N_V = 2$. While for $N_V = 1$ and $\gamma = 0.1$, the average residual tail energy is about -20.0 dBc, which could be a considerable interference when overlapping with adjacent packets. Base on the tail-cancelation performance and due to the need of complexity reduction, we only consider the following three schemes in the following simulations:

- $N_V = 2$ and overlapping the residual tails with adjacent packets;
- $N_V = 1$ and hard truncation; and
- $N_V = 1$ and truncation with windowing.

Other points that worth noting (but not shown in Fig. 6.6) are (i) When no tail shortening is applied (equivalently, $\gamma \rightarrow \infty$), the average tail energy is about -13.0 dBc. (ii) When N_V is large enough, such as 6, and $\gamma = 0$, $\xi_1 = -40.4$ dBc, and $\xi_2 = -11.9$ dBc. This value of ξ_2, in particular, indicates the significance of factoring the cancelation signal energy in our design equations.

Error Performance Without Residual Tail Truncation

In this section, we simulate a transmitter that overlaps the residual tails within the adjacent packets, as shown in Fig. 6.5. This simulation aims to quantify how BER performance is affected by overlapping the residual tails with adjacent packets. The channel for each packet has a unit gain, and random phase that is independent for each packet. AWGN noise is applied and no guard time (see Fig. 6.5) between packets is provided. Fig. 6.7 shows the results with $N_V = 1$ and 2, $\gamma = 0.1$, 12 active subcarriers, 64-QAM modulation and PHYDAYS filter. Performance of the original OQAM/FBMC system without overlapping is also provided for comparison. As observed, overlapping the residual tails with adjacent packets causes no observable influence on the BER performance when $N_V = 2$, with the selected γ. In comparison, as also shown in Fig. 6.7, overlapped OQAM/FBMC packets without tail-shortening result in a significant degradation in BER.

Error Performance With Residual Tail Truncation

As discussed earlier, truncation of $s(n)$ beyond K_e incurs distortion to the edge data symbols. This distortion is quantified by looking at the EVM of data symbols at the beginning and the end (i.e., the edge symbols) of each burst. EVM is defined as

$$\text{EVM (dBc)} = \text{Mean}(|q - \hat{q}|^2)/\text{Mean}(|q|^2), \qquad (6.22)$$

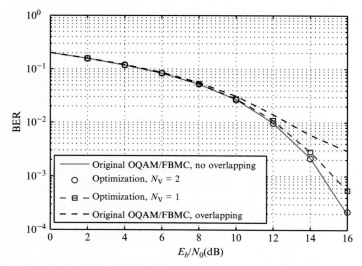

Fig. 6.7 BER performance without residual tail truncation, $N_V = 1$ and 2, $\gamma = 0.1$.

where q is a QAM symbol at the very edge of the burst, which consists of two PAM symbols, respectively, from the last and second last real-valued PAM OQAM/FBMC symbols, and $\hat{q}$ is the estimation of q at the receiver output, when the wireless channel is free of noise.

Fig. 6.8 plots EVM of the edge symbols for the optimization method with $N_V = 1$ and truncation, within a range of γ, for the case of PHYDAYS filter and 12 active subcarriers. Results of the hard truncation and truncation with windowing of the original OQAM/FBMC signal are presented for comparison. The tails are truncated at K_e for all the methods, therefore they all have the same tail overhead. The roll-off length L_r is set to $K/4$ for the truncation with windowing method. As results show, the optimization method with $N_V = 1$ lead to a significantly better EVM performance than the original OQAM/FBMC, for both the hard truncation and truncation with windowing. In particular, the EVM is improved from about -22 to -38 dBc by applying virtual symbols of $N_V = 1$ at $\gamma = 0.1$, for the truncation with windowing. It is also interesting to note that, with respect to EVM, hard truncation outperforms truncation with windowing. This is because, for a set value of K_e, windowing, in addition to hard truncation, adds some distortion to the first and last L_r samples of each burst.

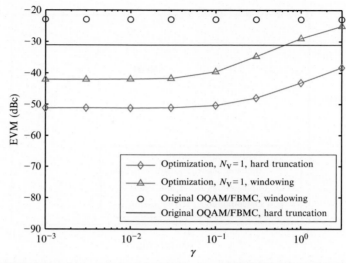

Fig. 6.8 EVM of the edge symbols with hard truncation and truncation with windowing, $N_V = 1$.

PAPR

PAPR is obtained by dividing the OQAM/FBMC signals into segments of length T, evaluating the peak power of each segment, and normalizing the peak power with respect to the average signal power. The curves of Fig. 6.9 then plot the Complementary Cumulative Distribution Function showing the probability of the PAPR exceeding a given PAPR threshold (PAPR$_0$).

Fig. 6.9 demonstrates the PAPR of the OQAM/FBMC bursts with the optimization method, for the case of PHYDAYS filter and 12 active subcarriers. PAPR of original OQAM/FBMC signal is also presented for comparison. As observed, the optimization method has a minimal effect on PAPR, when $\gamma = 0.1$.

PSD Performance

Fig. 6.10 shows the PSD of the optimization method with 12 active subcarriers. Here, we have adopted PHYDAS filter and the parameters $L_{\text{oh}} = K/4$ and $\gamma = 0.1$ are used. The performance of the original OQAM/FBMC signal with hard truncation and truncation with windowing are also presented for comparison. For the latter methods, the tail overhead is the same as the optimization method. Clearly, by

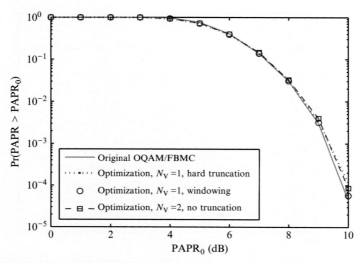

Fig. 6.9 PAPR of the OQAM/FBMC bursts with the optimization method, $\gamma = 0.1$.

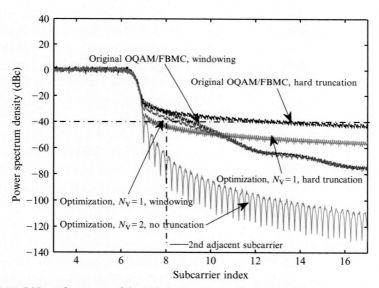

Fig. 6.10 PSD performance of the tail-shortening method with PHYDYAS filter, 12 active subcarrier, $L_{oh} = K/4$ and $\gamma = 0.1$.

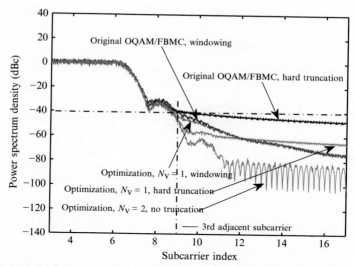

Fig. 6.11 PSD performance of the tail-shortening method with IOTA filter, 12 active subcarrier, $L_{oh} = K/4$ and $\gamma = 0.1$.

construction, the constructed waveforms according to the optimization method have exactly the same PSD as the original OQAM/FBMC signal, if the residual tail is not truncated (see the curve with $N_V = 2$). As results show, the optimization method with $N_V = 1$ considerably improves the PSD performance than the original OQAM/FBMC, for both the hard truncation and truncation with windowing.

For IOTA filter, the PSD of the optimization method with 12 active subcarriers, $\gamma = 0.1$ and $L_{oh} = K/4$, is presented in Fig. 6.11. Due to better concentration in time than PHYDYAS pulse, OQAM/FBMC signal with IOTA pulse has a sharper transition at the tail, however, higher OOB emission at adjacent subcarriers. Still, PSD performance with tail truncation is improved by the optimization method.

In Fig. 6.12, the PSD of the optimization method with 192 active subcarriers, $\gamma = 0.1$, $L_{oh} = K/4$, and PHYDYAS filter is presented. The 192 subcarriers are divided into 16 groups each with 12 subcarriers, and the optimization method is carried out within each group. This arrangement helps to reduce the computational complexity by a factor 16. It is not astonishing to find from Fig. 6.12 that the optimization method demonstrates almost the same performance as it does for 12 subcarriers in Fig. 6.10.

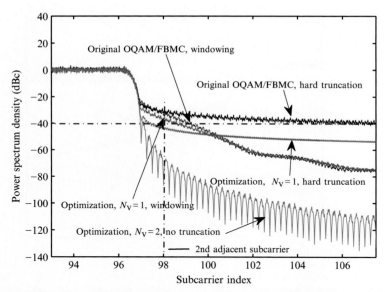

Fig. 6.12 PSD performance of the tail-shortening method with PHYDYAS filter, 192 active subcarrier, $L_{oh} = K/4$ and $\gamma = 0.1$.

Taking the error performance, PSD performance, and complexity into consideration, it is concluded that:

- The scheme with $N_V = 2$ and overlapping the residual tails with adjacent packets gives almost ideal error and PSD performance, however, a complexity higher than those with $N_V = 1$.
- The scheme with $N_V = 1$ and hard truncation has the lowest complexity and acceptable error performance; however, its OOB spectrum suppression is barely below -50 dBc.
- The scheme with $N_V = 1$ and truncation with windowing has low complexity and acceptable error performance, and a better OOB spectrum roll-off than that with hard truncation.

Comparison With UFMC and GFDM/CP-OQAM/FBMC

To complete our study and position the modified OQAM/FBMC waveform against recently proposed candidate waveforms for future wireless communications, we compare modified OQAM/FBMC with UFMC, GFDM, and CP-OQAM/FBMC waveforms in terms of their OOB emission and the packet length overhead incurred due to relevant filtering operations. Noting that GFDM and CP-OQAM/FBMC have a

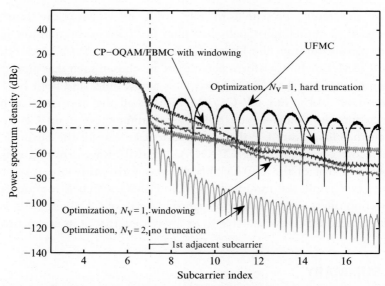

Fig. 6.13 PSD performance comparison among different multicarrier signals, 12 active subcarrier, $L_{oh} = K/4$ and $\gamma = 0.1$.

very similar OOB emission, we only present the PSD results of CP-OQAM/FBMC as a representative of the two.

Fig. 6.13 presents a set of PSD plots of UFMC, CP-OQAM/FBMC, and the OQAM/FBMC with virtual symbols. For UFMC, we have used the Dolph–Chebyshev filter with a stopband attenuation of 40 dB. In the literature, this choice has been widely advertised as the best compromised choice [17, 18]. The FFT length in UFMC is set equal to 256 and the length of Dolph–Chebyshev filter is equal to 19. As seen, due to the short length of the Dolph-Chebyshev filter, UFMC has a poor OOB emission for many subcarriers before it reaches the set 40 dB attenuation. Such attenuation is achieved only at subcarriers that are more 12 subcarriers away from the last active subcarrier in the passband of the waveform. The CP-OQAM/FBMC has a better OOB emission performance; however, still it does not reach the 40 dB attenuation until the fourth subcarrier away from the last active subcarrier in the passband. The OQAM/FBMC with $N_V = 2$ and no tail truncation, on the other hand, easily drops below the 40 dB attenuation line at the center of the first subcarrier adjacent to the passband. With $N_V = 1$ and tail truncation, the OQAM/FBMC still significantly outperforms UFMC and CP-OQAM/FBMC in terms of the frequency at which the OOB emission drops below −40 dBc.

For the results presented in Fig. 6.13, in UFMC, the overhead introduced by the Dolph-Chebyshev filter is 7.03% of the transmit signal before passing through the Dolph-Chebyshev filter. CP-OQAM/FBMC and OQAM/FBMC waveform parameters are chosen such that both carry the same length of data (7 QAM symbols across time) and both have tail overhead of 7.14%. For CP-OQAM/FBMC, all the overhead has been used to introduce raise-cosine roll-offs, meaning there is no CP in the waveform. Hence, any CP that is added to take care of the channel will increase the overhead of the CP-OQAM/FBMC waveform and make it longer than that of the OQAM/FBMC waveform with virtual symbols. We see that while the modified OQAM/FBMC which is introduced in this section offers a superior performance in terms of OOB emission, it has a tail overhead that is comparable to or lower than that of its competitors.

6.4 SUMMARY

This chapter introduced a method based on virtual symbols for tail shortening of OQAM/FBMC signals. The tail-shortening method transmits a set of virtual symbols at the two sides of each data packet to suppress signal samples at the OQAM/FBMC tails. Then, an optimization procedure was formulated to minimize the tail energy beyond a set sample index. The results showed that while the optimization method retains the excellent OOB emission performance of the conventional OQAM/FBMC method, a very low EVM, and a very minimal change in PAPR, it results in a tail overhead that remains comparable to or lower than those of the UFMC, GFDM and CP-OQAM/FBMC, which in recent literature have been introduced as its competitors.

REFERENCES

[1] Farhang-Boroujeny B. OFDM versus filter bank multicarrier. IEEE Signal Process Mag 2011;28(3):92–112.
[2] Siohan P, Siclet C, Lacaille N. Analysis and design of OQAM-OFDM systems based on filterbank theory. IEEE Trans Signal Process 2002;50(5):1170–83.
[3] Gao X, Wang W, Xia XG, Au EKS, You X. Cyclic prefixed OQAM-OFDM and its application to single-carrier FDMA. IEEE Trans Commun 2011;59(5):1467–80.
[4] Hirosaki B. An orthogonally multiplexed QAM system using the discrete Fourier transform. IEEE Trans Commun 1981;29(7):982–9.
[5] Chen D, Qu D, Jiang T. Prototype filter optimization to minimize stopband energy with NPR constraint for filter bank multicarrier modulation systems. IEEE Trans Signal Process 2013;61(1):159–69.

[6] Zhang H, Ruyet DL, Roviras D, Sun H. Noncooperative multicell resource allocation of FBMC-based cognitive radio systems. IEEE Trans Veh Technol 2012;61(2):799–811.

[7] Ihalainen T, Ikhlef A, Louveaux J, Renfors M. Channel equalization for multi-antenna FBMC/OQAM receivers. IEEE Trans Veh Technol 2011;60(5):2070–85.

[8] Chang R, Gibbey R. A theoretical study of performance of an orthogonal multiplexing data transmission scheme. IEEE Trans Commun Technol 1968;16(4):529–40.

[9] Zou WY, Wu Y. COFDM: an overview. IEEE Trans Broadcast 1995;41(1):1–8.

[10] Bellanger M. Efficiency of filter bank multicarrier techniques in burst radio transmission. In: Global telecommunications conference; 2010.

[11] Bellanger M, Renfors M, Ihalainen T, Rocha CA. OFDM and FBMC transmission techniques: a compatible high performance proposal for broadband power line communications. In: International symposium on power line communications and its applications; 2010.

[12] Abdoli MJ, Jia M, Ma J. Weighted circularly convolved filtering in OFDM/OQAM. In: 24th International symposium on personal indoor and mobile radio communications, London, UK; 2013.

[13] Bellanger MG. Specification and design of a prototype filter for filter bank based multicarrier transmission. In: IEEE international conference on acoustics, speech, and signal processing; 2001.

[14] Mirabbasi S, Martin K. Overlapped complex-modulated transmultiplexer filters with simplified design and superior stopbands. IEEE Trans Circuits Syst II Analog Digit Signal Process 2003;50(8):456–69.

[15] Viholainen A, Bellanger M, Huchard M. Prototype filter and structure optimization, 2009. Available from: http://www.ict-phydyas.org/delivrables/PHYDYAS-D5-1.pdf/view.

[16] Dahlman E, Parkvall S, Skold J. 4G: LTE/LTE-advanced for mobile broadband. London: Academic Press; 2013.

[17] Wunder G, Jung P, Kasparick M, et al. 5GNOW: non-orthogonal, asynchronous waveforms for future mobile applications. IEEE Commun Mag 2014;52(2):97–105.

[18] Wang X, Wild T, Schaich F. Filter optimization for carrier- frequency- and timing-offset in universal filtered multi-carrier systems. In: IEEE vehicular technology conference; 2015.

CHAPTER 7

Synchronization

Compared with single-carrier systems, multicarrier systems are more sensitive to synchronization offset. The orthogonality among the subcarriers usually depends on the carrier frequency offset (CFO) as well as the time offset (TO). When multicarrier systems suffer from the synchronization offset problem, the orthogonality among the subcarriers is damaged, resulting in severe intersymbol interference (ISI) to data symbols. Hence, synchronization methods are necessary for multicarrier systems. Of course, no matter how precise the synchronization algorithm is, the synchronization offset is difficult to be completely eliminated, since the local oscillator offset and the Doppler shift always exist in practice.

As a multicarrier system, offset quadrature amplitude modulation-based filter bank multicarrier (OQAM/FBMC) also suffers the synchronization offset problem. Hence, accurate and reliable TO and CFO synchronization schemes must be designed for OQAM/FBMC systems. Among many synchronization methods, blind synchronization and data-aided methods are two major techniques. The blind synchronization methods can achieve good performance without additional training sequences. However, the blind synchronization methods usually take a long time to estimate TO and CFO, due to the high computational complexity. Currently, the properties of the transmitted OQAM/FBMC signal have been exploited in order to obtain blind CFO estimators in [1–4]. On the other hand, data-aided synchronization methods can obtain the synchronization offsets in a faster way by designing some special training sequences. Some data-aided synchronization schemes have been proposed in [5–8] for OQAM/FBMC systems. In this chapter, traditional data-aided synchronization methods, such as TR1, TR2, LS (MLS), etc., are introduced in detail. Then, a new data-aided timing and CFO synchronization scheme is presented by exploiting the conjugate-symmetry property of a properly designed training sequence. Finally, this chapter describes a case of Bayesian CFO estimation as an example of blind synchronization methods.

The rest of this chapter is organized as follows. Section 7.1 describes the effects of timing and frequency offsets in OQAM/FBMC. The traditional data-aided synchronization schemes proposed in [5] are introduced in Section 7.2. In Section 7.3, a new data-aided timing and CFO synchronization scheme is presented by exploiting the conjugate-symmetry property of a properly designed training sequence. Section 7.4 describes a case of Bayesian CFO estimation. Finally, summary is made in Section 7.5.

7.1 EFFECTS OF TIMING AND FREQUENCY OFFSETS

Synchronization is required at the receiver to compensate for the CFO between the incoming signal and the local oscillator, as well as the TO between the transmitter and receiver [9]. CFO destroys orthogonality among subcarriers and results in intercarrier interference (ICI) as well as multiple access interference. TO results in ISI and must be counteracted to avoid performance degradation.

The baseband OQAM/FBMC signal is expressed as

$$s(t) = \sum_{k=0}^{K-1} \sum_{m \in Z} a_k(m) \underbrace{h(t - m\tau_0)\, e^{j2\pi k F_0 t} e^{j\phi_{k,m}}}_{h_{k,m}(t)}, \tag{7.1}$$

where F_0 is the intercarrier spacing, τ_0 is the real-valued symbol duration ($\tau_0 = 1/2F_0$), $h(t)$ is the prototype function, and K is the number of subcarriers.

The general channel model can be expressed as

$$g(t, \tau) = \sum_{i=0}^{P-1} c_i e^{j2\pi f_d^i t} \delta(\tau - d_i), \tag{7.2}$$

where P is the number of paths (the first path is the reference path with delay $d_0 = 0$) and c_i is the channel gain of the ith path which can be expressed as $c_i = \rho_i e^{j\theta_i}$, with ρ_i the attenuation of the ith path and θ_i the phase rotation due to the delay d_i, f_d^i is the Doppler frequency of the ith path. $\delta(\cdot)$ is the Dirac delta function.

After passing through the channel, the received OQAM/FBMC signal can be expressed as

$$y(t) = \sum_{i=0}^{P-1} g(t, d_i) s(t - d_i)$$

$$= \sum_{i=0}^{P-1} c_i e^{j2\pi f_d^i t} \times \sum_{m \in Z} \sum_{k=0}^{K-1} a_k(m) h(t - d_i - m\tau_0) e^{j2\pi k F_0(t-d_i)} e^{j\phi_{k,m}}.$$

$$(7.3)$$

The demodulated OQAM/FBMC signal, for the frequency-time position (k_0, m_0), is obtained by applying a matched filter yielding [10]

$$y_{k_0}(m_0) = \int_{-\infty}^{+\infty} y(t) g_{k_0, m_0}^{\star}(t) dt \qquad (7.4)$$

$$= \sum_{m \in Z} \sum_{k=0}^{K-1} a_k(m) e^{j\phi_1} \sum_{i=0}^{P-1} c_i e^{j\phi_2} \times A_g((m_0 - m)\tau_0 - d_i, (k_0 - k)F_0 - f_d^i),$$

where $\phi_1 = \phi_{k,m} - \phi_{k_0,m_0}$ and ϕ_2 is given as

$$\phi_2 = 2\pi \left(\left((k_0 - k)F_0 + f_d^i \right) \left(\frac{m_0 + m}{2} \tau_0 + \frac{d_i}{2} \right) - k F_0 d_i \right). \qquad (7.5)$$

The ambiguity function A_g is defined as

$$A_g(\tau, \nu) = \int_{-\infty}^{+\infty} h\left(t + \frac{\tau}{2}\right) h^{\star}\left(t - \frac{\tau}{2}\right) e^{-j2\pi \nu t} dt. \qquad (7.6)$$

The discrete-time format of Eq. (7.4) is given as

$$y_{k_0}(m_0) = \sum_{(p,q)} a_{k_0+p}(m_0 + q) e^{j(\pi/2)(p+q+pq)} e^{j\pi p m_0}$$

$$\times \sum_{l=0}^{L_h-1} c_l e^{j\pi((2m_0+q)/2 + l/K)(1/r)}$$

$$\times A_g[-qK/2 - l, -(p + 1/r)] e^{-j\pi(2k_0+p)l/K}, \qquad (7.7)$$

where $k = k_0 + p$, $m = m_0 + q$, $F_0 = 1/KT_s$, and $A_g[q, p] = A(qT_s, pF_0)$. The parameter $r = \frac{1}{f_d T_0}$ is related to the Doppler shift.

The demodulated signal can then be divided into the signal part and the interference part as

$$y_{k_0}(m_0) = \alpha_{k_0}(m_0) a_{k_0}(m_0) + J_{k_0}(m_0), \qquad (7.8)$$

with

$$\alpha_{k_0}(m_0) = \sum_{l=0}^{L_h-1} c_l e^{j\pi(m_0+l/K)(1/r)} A_g[-l, -1/r] e^{-j\pi 2k_0 l/K}, \quad (7.9)$$

and

$$J_{k_0}(m_0) = \sum_{(p,q)\neq(0,0)} a_{k_0+p}(m_0 + q) e^{j(\pi/2)(p+q+pq)} e^{j\pi p m_0}$$

$$\times \sum_{l=0}^{L_h-1} c_l e^{j\pi((2m_0+q)/2+l/K)(1/r)} \times A_g[-qK/2 - l,$$

$$- (p + 1/r)] e^{-j\pi(2k_0+p)l/K}. \quad (7.10)$$

7.1.1 Impacts of TO

For simplicity, assume a simple one-path delayed model, that is,

$$g(\tau) = \delta(\tau - d_0). \quad (7.11)$$

Hence, Eqs. (7.9), (7.10) are rewritten as

$$\alpha_{k_0} = A_g[-l_d, 0] e^{-j\pi 2k_0 l_d/K}, \quad (7.12)$$

$$J_{k_0}(m_0) = \sum_{(p,q)\neq(0,0)} a_{k_0+p}(m_0 + q) e^{j(\pi/2)(p+q+pq)} e^{j\pi p m_0}$$

$$\times \underbrace{A_g[-qK/2 - l_d, -p] e^{-j\pi(2k_0+p)l_d/K}}_{T_{k_0}^{(p,q)}}. \quad (7.13)$$

Since the channel is now time-invariant, the channel coefficient α_{k_0} and $T_{k_0}^{(p,q)}$ are no longer time dependent.

When a simple one-tap zero-forcing (ZF) equalizer is employed, the signal after the real part extraction operation is given as

$$\hat{a}_{k_0}(m_0) = \text{Re}\left\{\frac{\alpha_{k_0}}{H'_{k_0}}\right\} a_{k_0}(m_0) + \text{Re}\left\{\frac{\alpha_{k_0}}{H'_{k_0}}\right\} \quad (7.14)$$

$$= \text{Re}\left\{\alpha_{k_0}^{\{ZF\}}\right\} a_{k_0}(m_0) + \text{Re}\left\{J_{k_0,m_0}^{\{ZF\}}\right\},$$

where H'_{k_0} is the ZF equalizer coefficient at the k_0th subcarrier.

The interference, before the real part extraction, can be written as

$$J_{k_0,m_0}^{\{ZF\}} = \sum_{(p,q)\neq(0,0)} a_{k_0+p}(m_0+q)e^{j(\pi/2)(p+q+pq)+\pi(pm_0-pl_d/K)}$$

$$A_g[-qK/2-l_d,-p]. \tag{7.15}$$

Note that the symbols are zero-mean variables, hence, the interference power can be obtained as

$$P_J^{\mathbf{Re},ZF}(l_d) = \mathbf{E}\left[\left|\mathbf{Re}\left\{J_{k_0,m_0}^{\{ZF\}}\right\}\right|^2\right] \tag{7.16}$$

$$= \sigma_a^2 \sum_{(p^0,q^0)} \cos^2\left(\frac{\pi}{2}(p+q+pq)-\pi p\frac{l_d}{K}\right) \times A_g^2[-qK/2-l_d,-p],$$

where $\mathbf{E}[\cdot]$ denotes the expectation operator and σ_a^2 is the variance of the real OQAM/FBMC symbols. Thus, the signal-to-interference ratio (SIR) expression, ratio of the useful signal and interference powers, yields [11]

$$\text{SIR}(l_d) = \frac{A_g^2[-l_d,0]}{\sum_{(p,q)\neq(0,0)}\cos^2\left(\frac{\pi}{2}(p+q+pq)-\pi p\frac{l_d}{K}\right)A_g^2[-qK/2-l_d,-p]}. \tag{7.17}$$

7.1.2 Impacts of CFO

Consider a simple case where the CFO is assumed to be static. The channel model can then be expressed as

$$g(t,\tau) = e^{2\pi f_d t}\delta(\tau).$$

Thus, the discrete-time demodulated signal at position (k_0, m_0) is given as

$$y_{k_0}(m_0) = \underbrace{e^{j\pi m_0(1/r)}A_g[0,-1/r]}_{\alpha_{m_0}}a_{k_0}(m_0) + J_{k_0}(m_0), \tag{7.18}$$

with

$$J_{k_0}(m_0) = \sum_{(p,q)\neq(0,0)} a_{k_0+p}(m_0+q)e^{j(\pi/2)(p+q+pq)}e^{j\pi pm_0}$$

$$\times \underbrace{e^{j\pi(2m_0+q)/(2r)}A_g[-qK/2,-(p+1/r)]}_{T_{m_0}^{(p,q)}}. \tag{7.19}$$

Now the channel coefficients α_{m_0} and $T_{m_0}^{(p,q)}$ are no longer frequency dependent.

By taking into account the ZF equalizer with coefficient $H'_{m_0} = e^{j\pi m_0(1/r)}$, the interference power after real part extraction is given as

$$P_J^{\text{Re},ZF}(1/r) = \mathbf{E}\left[\left|\mathbf{Re}\left\{\frac{J_{k_0}(m_0)}{H'_{m_0}}\right\}\right|\right]$$

$$= \sigma_a^2 \sum_{(p^0,q^0)} \cos^2\left(\frac{\pi}{2}\left(p+q+pq+q/r\right)\right)$$

$$\times A_g^2[-qK/2, -(p+1/r)]. \qquad (7.20)$$

Finally, the analytical SIR, at CFO $= 1/r$, is given as [11]

$$\text{SIR}(1/r) = \frac{A_g^2[0,-1/r]}{\sum_{(p,q)\neq(0,0)} \cos^2\left(\frac{\pi}{2}\left(p+q+pq+q/r\right)\right)A_g^2[-qK/2, -(p+1/r)]}. \qquad (7.21)$$

7.2 DATA-AIDED SYNCHRONIZATION SCHEMES WITH IDENTICAL PROPERTY

As one of multicarrier systems, OQAM/FBMC also has synchronization offset problem. Hence, accurate and reliable TO and CFO synchronization schemes must be designed for OQAM/FBMC systems. Among many synchronization methods, data–aided and blind synchronization methods are two major techniques. The data–aided joint CFO and symbol timing estimator based on the LS approach is introduced in this section [5].

This section considers an OQAM/FBMC system with K subcarriers, where only K_u subcarriers are used to transmit data. The transmitted signal is given as

$$s(nT_s) = \sqrt{\frac{K}{2K_u}} \sum_{p=-\infty}^{\infty} \sum_{l\in\mathcal{A}} e^{jl\left(\frac{2\pi}{T}nT_s+\frac{\pi}{2}\right)}$$

$$\times [a_l^R(p)h(nT_s - pKT_s) + ja_l^{I,TR}h(nT_s - KT_s/2 - pKT_s)], \qquad (7.22)$$

where $a_l^R(p)$ and $a_l^I(p)$ are the real and imaginary parts of the pth transmitted complex data, respectively. $\mathcal{A}$ denotes the set of indices of data subcarriers, T is the signaling interval, and $T_s = T/K$. After passing through the channel, the received signal $r(t)$ is given as

$$r(nT_s) = e^{j(2\pi \Delta fT_s n+\phi)}s(nT_s - \tau) + \mathcal{N}(nT_s), \qquad (7.23)$$

where Δf is the CFO, ϕ is the carrier phase offset, and τ is the TO. In addition, $\mathcal{N}(nT_s)$ denotes the noise.

By constructing some identical blocks in the training sequence, the receiver can obtain the timing estimation by looking for the maximum correlation output. For example, the training sequence can be obtained by transmitting data symbols $a_l(p) = a_l^{TR}$ for $\forall l \in \mathcal{A}$ and $\forall p \in \{0, \ldots, N_{TR} - 1\}$. Assume that the prototype filter $h(t)$ is not zero for $t \in \{0, T_s, \ldots, (N_g - 1)T_s\}$, where $N_g = \gamma T/T_s$ with γ the overlapping factor. The training burst is then obtained as

$$s_{TR}(nT_s) = \sqrt{\frac{K}{2K_u}} \sum_{p=0}^{N_{TR}-1} \sum_{l \in \mathcal{A}} e^{jl\left(\frac{2\pi}{K}n + \frac{\pi}{2}\right)}$$
$$\times [a_l^{R,TR} h(nT_s - pKT_s) + ja_l^{I,TR} h(nT_s - KT_s/2 - pKT_s)]. \tag{7.24}$$

After a transient of $N_a - 1$ samples, in particular for $n \in \{N_a - 1, \ldots, N_{TR}K - P - 1\}$, the training sequence in Eq. (7.24) satisfies the following relationship:

$$s_{TR}(nT_s + PT_s) = s_{TR}(nT_s), \tag{7.25}$$

with $P = K$. The relationship only holds for the middle samples due to the prototype filter employed in OQAM/FBMC systems. Thus, the number of identical blocks contained in the training sequence is $N_{rip} = N_{TR} - \gamma$.

Accounting for the relationship in Eq. (7.25), a joint symbol timing and CFO estimator can be obtained by considering the minimization problem

$$\left(\Delta \hat{f}, \hat{\tau}\right) = \arg \min_{\Delta \hat{f}, \hat{\tau}} \left\{ \sum_{k=N_a-1}^{N_{TR}K-P-1} \left| r(nT_s + \tilde{\tau}) - r(nT_s + PT_s + \tilde{\tau}) e^{-j2\pi\Delta\hat{f}T_sP} \right|^2 \right\}, \tag{7.26}$$

where $\Delta \hat{f}$ and $\hat{\tau}$ are trial values for CFO and symbol timing, respectively. The minimization in Eq. (7.26) leads to the following joint CFO and symbol timing estimator referred to as LS estimator:

$$\hat{\tau}_{LS} = \arg \max_{\tilde{\tau}} \left\{ 2 \left| R(\tilde{\tau}) \right| - Q_1(\tilde{\tau}) - Q_2(\tilde{\tau}) \right\}, \tag{7.27}$$

$$\Delta \hat{f}_{LS}(\hat{\tau}_{LS}) = \frac{1}{2\pi P T_s} \angle \left\{ R(\hat{\tau}_{LS}) \right\}, \tag{7.28}$$

with

$$R(\tilde{\tau}) \triangleq \sum_{k=N_a-1}^{N_{TR}K-P-1} r^*(nT_s + \tilde{\tau}) r(nT_s + PT_s + \tilde{\tau}), \tag{7.29}$$

and

$$Q_i(\tilde{\tau}) \triangleq \sum_{k=N_a-1}^{N_{TR}K-P-1} |r(nT_s + (i-1)PT_s + \tilde{\tau})|^2, \qquad i = 1, 2. \quad (7.30)$$

If the timing metric in Eq. (7.27) is divided by the term $Q(\tilde{\tau}) \triangleq Q_1(\tilde{\tau}) + Q_2(\tilde{\tau})$, a modified LS (MLS) [5] estimator is obtained as

$$\hat{\tau}_{MLS} = \arg \max_{\tilde{\tau}} \left\{ \frac{|R(\tilde{\tau})|}{Q(\tilde{\tau})} \right\}, \qquad (7.31)$$

$$\Delta\hat{f}_{MLS}(\hat{\tau}_{MLS}) = \frac{1}{2\pi PT_s} \angle \left\{ R(\hat{\tau}_{MLS}) \right\}. \qquad (7.32)$$

In addition, Tonello and Rossi proposed two joint estimators in [6] referred as TR1 and TR2. The TR1 estimator is expressed as

$$\hat{\tau}_{TR1} = \arg \max_{\tilde{\tau}} \left\{ \frac{|R(\tilde{\tau})|^2}{Q(\tilde{\tau})^2} \right\}, \qquad (7.33)$$

$$\Delta\hat{f}_{TR1}(\hat{\tau}_{TR1}) = \frac{1}{2\pi PT_s} \angle \left\{ R(\hat{\tau}_{TR1}) \right\}. \qquad (7.34)$$

The TR2 estimator is expressed as

$$\hat{\tau}_{TR2} = \arg \max_{\tilde{\tau}} \left\{ \frac{|S(\tilde{\tau})|^2}{T(\tilde{\tau})^2} \right\}, \qquad (7.35)$$

$$\Delta\hat{f}_{TR2}(\hat{\tau}_{TR2}) = \frac{1}{2\pi PT_s} \angle \left\{ S(\hat{\tau}_{TR2}) \right\}, \qquad (7.36)$$

where

$$S(\tilde{\tau}) = \sum_{k=N_a-1}^{N_{TR}K-P-1} \left[r^\star(nT_s+\tilde{\tau})\, s_{TR}(nT_s)\, r(nT_s+PT_s+\tilde{\tau})\, s_{TR}^\star(nT_s+PT_s) \right],$$
$$(7.37)$$

and

$$T(\tilde{\tau}) = \sum_{k=N_a-1}^{N_{TR}K-P-1} |s_{TR}(nT_s)|^2 |s_{TR}(nT_s + PT_s)|^2. \qquad (7.38)$$

The TR2 algorithm using the cross correlation of the received signal is superior to the TR1 and MLS algorithms using the autocorrelation in terms of the TO estimation. On the other hand, TR1 and MLS algorithms are superior to the TR2 algorithm in terms of the CFO estimation.

7.3 DATA-AIDED SYNCHRONIZATION SCHEME WITH CONJUGATE-SYMMETRY PROPERTY

In this section, we present a method to design training sequence with conjugate-symmetry property and introduce the corresponding timing offset estimation algorithm and CFO estimation algorithm based on the training sequence, respectively.

7.3.1 Design of Training Sequence

First, this subsection aims to obtain a properly designed training sequence satisfying conjugate-symmetry property, which is similar to that in OFDM systems, namely $s(A - B) = s^*(A + B)$, where A is the conjugate-symmetry center and B is the width. Specially, one pilot symbol is transmitted every fourth subcarrier at the n_0th instant, that is, on the $(m = 4m_0)$th subcarrier with $m_0 = 0, 1, \ldots, K/4 - 1$, the sequence can be obtained as

$$
s_0(k) = \sum_{m_0=0}^{\frac{K}{4}-1} a_{4m_0}(n_0) h\left(k - \frac{n_0 K}{2}\right) e^{\frac{j2\pi \cdot 4m_0 \cdot k}{K}} e^{\frac{j(4m_0 + n_0)\pi}{2}}
$$

$$
= \sum_{m_0=0}^{\frac{K}{4}-1} a_{4m_0}(n_0) h\left(k - \frac{n_0 K}{2}\right) e^{\frac{j2\pi \cdot 4m_0 \cdot k}{K}} e^{\frac{j\pi n_0}{2}}. \tag{7.39}
$$

In general, $h(k)$ is a real-valued symmetrical pulse-shaping filter with the length of αK, that is, $h(k) = h(\alpha K - k)$. Then, it can be obtained as

$$
h\left(k - \frac{n_0 K}{2}\right) = h\left(\alpha K - k + \frac{n_0 K}{2}\right). \tag{7.40}
$$

By substituting $k = \alpha K + n_0 K - k$ to Eq. (7.39), it can be written as

$$
s_0(\alpha K + n_0 K - k) = \sum_{m_0=0}^{\frac{K}{4}-1} a_{4m_0}(n_0) h\left(\alpha K + \frac{n_0 K}{2} - k\right) e^{\frac{j2\pi \cdot 4m_0 \cdot (-k)}{K}} e^{\frac{j\pi n_0}{2}}. \tag{7.41}
$$

For simplification, assume that n_0 is odd, then $e^{\frac{j\pi n_0}{2}} = \pm 1$. Comparing previous Eqs. (7.39)–(7.41), we have $s_0(k) = s_0^*(\alpha K + n_0 K - k)$. To easily illustrate the symmetry of training sequence, for $k = 0, 1, \ldots, N/2 - 1$, we have

$$
s_0\left(\frac{\alpha K + n_0 K}{2} - k\right) = s_0^*\left(\frac{\alpha K + n_0 K}{2} + k\right). \tag{7.42}
$$

Obviously, the training sequence is conjugate-symmetry with respect to $k = \frac{\alpha K + n_0 K}{2}$, which is suitable for the TO and CFO estimations.

However, due to the overlapping structure of the adjacent OQAM/F-BMC symbols, we should separate the pilot symbols and data symbols by inserting zero-valued symbols. When the number of zero-valued symbols is larger, the interference to pilot symbols is smaller. However, the insertion of zero-valued symbols reduces spectrum efficiency. Therefore, it is interesting to investigate how to achieve good synchronization performance while keeping the number of zero-valued symbols as small as possible meanwhile.

7.3.2 TO Estimation

According to the previous training sequence, we present a TO estimator as follows. First, denote

$$R(\tilde{\tau}) = \sum_{k=0}^{K/2-1} r\left(\frac{\alpha K + n_0 K}{2} - k + \tilde{\tau}\right) r\left(\frac{\alpha K + n_0 K}{2} + k + \tilde{\tau}\right). \quad (7.43)$$

We can then obtain

$$
\begin{aligned}
R(\tilde{\tau}) = \sum_{k=1}^{K/2-1} & s\left(\frac{\alpha K + n_0 K}{2} - k + \tilde{\tau} - \tau\right) \\
& s\left(\frac{\alpha K + n_0 K}{2} + k + \tilde{\tau} - \tau\right) e^{j2\pi \frac{2\left(\frac{\alpha K + n_0 K}{2} + \tilde{\tau}\right)\Delta f}{K}} \\
& + s\left(\frac{\alpha K + n_0 K}{2} - k + \tilde{\tau} - \tau\right) e^{j2\pi \frac{\left(\frac{\alpha K + n_0 K}{2} + \tilde{\tau} - k\right)\Delta f}{K}} n\left(\frac{\alpha K + n_0 K}{2} + k + \tilde{\tau}\right) \\
& + s\left(\frac{\alpha K + n_0 K}{2} + k + \tilde{\tau} - \tau\right) e^{j2\pi \frac{\left(\frac{\alpha K + n_0 K}{2} + \tilde{\tau} + k\right)\Delta f}{K}} n\left(\frac{\alpha K + n_0 K}{2} - k + \tilde{\tau}\right) \\
& + n\left(\frac{\alpha K + n_0 K}{2} + k + \tilde{\tau}\right) n\left(\frac{\alpha K + n_0 K}{2} - k + \tilde{\tau}\right).
\end{aligned}
\quad (7.44)
$$

In Eq. (7.44), four terms can be obtained, including one signal term, one noise term, and two cross terms. In high SNR condition, the power of noise $w(n)$ is much lower than signal power, one can ignore the noise and cross terms for simplicity first. When the estimated timing offset equals the real value, that is, $\tilde{\tau} = \tau$, according to the conjugate-symmetry property, it can be obtained by

$$R(\tau) \approx \sum_{k=1}^{K/2-1} \left| s\left(\frac{\alpha K + n_0 K}{2} + k\right) \right|^2 e^{j\frac{4\pi\left(\frac{\alpha K + n_0 K}{2} + \tau\right)\Delta f}{K}}. \quad (7.45)$$

In addition, we define

$$Q(\tilde{\tau}) = \max \sum_{k=0}^{K/2-1} \left\{ \left| r\left(\frac{\alpha K + n_0 K}{2} - k + \tilde{\tau}\right)\right|^2, \left| r\left(\frac{\alpha K + n_0 K}{2} + k + \tilde{\tau}\right)\right|^2 \right\}.$$

(7.46)

To normalize $R(\tilde{\tau})$, the final timing metric function can be written as

$$M(\tilde{\tau}) = \frac{|R(\tilde{\tau})|^2}{Q(\tilde{\tau})^2}.$$

(7.47)

Therefore, the estimated timing offset can be obtained as

$$\hat{\tau} = \arg\max_{\tilde{\tau}} M(\tilde{\tau}).$$

(7.48)

When $\tilde{\tau} = \tau$, $R(\tilde{\tau})$ is the linear in-phase superposition of $K/2 - 1$ samples and $M(\tilde{\tau})$ should be the largest. When $\tilde{\tau} \neq \tau$, $R(\tilde{\tau})$ is the nonlinear superposition of $K/2 - 1$ samples, and it is obvious smaller than that when $\tilde{\tau} = \tau$. Therefore, the corresponding $M(\tilde{\tau})$ can be regarded as noise. In conclusion, due to the conjugate-symmetry property, $M(\tilde{\tau})$ is very sharp for easily detection.

Moreover, considering the fact that the first path is not always the instantaneous power strongest path in actual multipath fading channel, one should combine the earlier estimation algorithm with sliding window to improve the accuracy of the timing synchronization. The core idea of the sliding window algorithm is to apply the final timing metric function $M(\tilde{\tau})$ to a sliding window with fixed length D ($D > \tau_{max}$) to obtain a new timing metric $M_D(\tilde{\tau})$, where τ_{max} is the maximum delay of the multipath channel. The timing estimator is then obtained as

$$\hat{\tau} = \arg\max_{\tilde{\tau}} \sum_{m=0}^{D-1} M(\tilde{\tau} + m)$$

$$= \arg\max_{\tilde{\tau}} M_D(\tilde{\tau}).$$

(7.49)

7.3.3 CFO Estimation

In order to achieve accurate CFO estimation with low implementation complexity, one should construct an identical part Syn2 after the above training symbol part Syn1.

After timing offset estimation, one should consider about the identical property of Syn1 and Syn2, and the following correlation can be obtained

$$R_1(\hat{\tau}) = \sum_{k=0}^{K-1} r^\star \left(\frac{\alpha K + n_0 K}{2} - \frac{K}{2} + k + \hat{\tau} \right) \cdot r \left(\frac{\alpha K + n_0 K}{2} - \frac{K}{2} + d + k + \hat{\tau} \right),$$

(7.50)

where d is interval samples number between Syn1 and Syn2.

As a result, the CFO can be estimated as

$$\Delta \hat{f}(\hat{\tau}) = \frac{K}{d} \cdot \frac{1}{2\pi} \angle \{ R_1(\hat{\tau}) \}.$$

(7.51)

7.3.4 Simulation Results

In this subsection, the performance of the synchronization algorithm is given via computer simulations in both AWGN and 802.22A channels, and the simulation parameters of OQAM/FBMC systems are shown in Table 7.1. The values of the CFO Δf normalized to subcarrier spacing and the timing offset τ are uniformly distributed in the range of $[-1/4, 1/4]$ and $\{-K/4, \ldots, K/4\}$, respectively.

The probability of exactly estimating timing offset per trial, P_catch, is used to evaluate the performance of the TO estimation. Usually, the probability can be statistically illustrated and can be defined as $P\_catch = N_t/N_{\text{total}}$, where N_{total} (large enough) is the number of TO estimation trials and N_t is the number of exact TO estimation. In multipath channels, the mean squared error of timing offset estimator is defined as $MSE\_STO = (\hat{\tau} - \tau)^2$. In addition, the mean squared error of CFO estimator is defined as $MSE\_CFO = (\Delta \hat{f} - \Delta f)^2$.

Table 7.1 Simulation parameters

System parameters		Value
Number of subcarriers (K)		256
Modulation scheme		QPSK
Prototype function		IOTA filters (length: 4K)
Number of pilot symbols		4.5
802.22A channel	Delay of channel (T_s)	[0 2 4 7 11 14]
	Gain of channel (dB)	[0 − 7 − 15 − 22 − 24 − 19]

Performance of Timing Offset Estimator

Fig. 7.1 shows *P_catch* of TO estimation versus SNR in AWGN and 802.22A channels. It is obvious that in AWGN channel, the probability remains 1 for all SNRs, which means that the algorithm can exactly estimate timing offset in per trial without errors. The estimator is unbiased and can achieve perfect performance of timing offset estimation in AWGN channel.

In 802.22A channel, with the increase of SNR, the probability grows rapidly, and especially when SNR $\geq$ 10 dB, it is up to a high value of 0.98. The timing estimator can reach almost the same performance with the window-based synchronization (WBS) scheme when SNR $\geq$ 10 dB, although a little but acceptable gap may exist in low SNR. However, in the WBS scheme, one should jointly consider timing offset estimation and CFO estimation at the same time, while the scheme can achieve the above synchronization independently in steps with relatively lower implementation complexity. Note that the probability cannot reach 1, which means the estimator is still difficult to reach perfect estimation.

Fig. 7.2 shows the mean squared error of timing estimator in 802.22A channel. When SNR $\geq$ 10 dB, MSE_STO is less than 10 degree, which means that the mean error of per-trial is lower than 1 sample. Therefore, the receiver needs to perform proper channel estimation to complete demodulation.

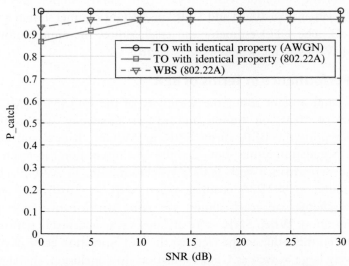

Fig. 7.1 Capture probability of timing synchronization in AWGN and 802.22A channels.

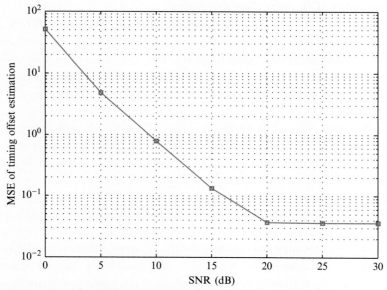

Fig. 7.2 MSE of timing offset estimation in 802.22A channel.

7.3.5 Performance of CFO Estimator

Fig. 7.3 presents the performance of the CFO estimator. Numerical results show that, with the increase of SNR, the mean square error decreases rapidly. For the condition MSE_CFO $< 10^{-5}$, SNR in 802.22A channel should be larger than 10 dB and SNR in AWGN channel should be larger than 7 dB. The CFO estimator can reach almost the same performance with the WBS scheme when SNR ≥ 10 dB.

7.3.6 Performance in Terms of BER

To gain some insight about the sensitivity of the OQAM/FBMC signal in terms of BER on timing and carrier frequency synchronization, Fig. 7.4 reports the BER versus SNR in AWGN channel and 802.22A channel. Simulation results show that the synchronization can reach almost the same performance with perfect synchronization in AWGN channel. Furthermore, in 802.22A channel, when SNR ≥ 10 dB, the synchronization scheme can reach almost the same performance with perfect synchronization. When SNR < 10 dB, there is an acceptable performance gap between them due to the fact that the TO estimation error of per-trial is relatively large as well as CFO estimation. In conclusion, the algorithm can achieve good synchronization, and improve the system performance.

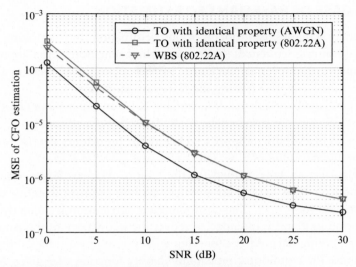

Fig. 7.3 MSE of CFO estimation in AWGN and 802.22A channel.

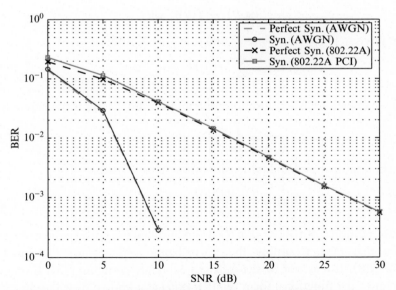

Fig. 7.4 BER performance of OQAM/FBMC system in AWGN and 802.22A channel.

7.4 A CASE OF BAYESIAN CFO ESTIMATION

Data–aided schemes can obtain robust estimation by exploiting well-designed training sequences. In contrast, blind synchronization approaches can improve the spectral efficiency. In this section, we present a blind synchronization approach using the Bayesian estimation.

Suppose an observable random variable $\mathbf{X}$ for an experiment takes values in the set S, and the distribution of $\mathbf{X}$ depends on a parameter θ, which takes values in a parameter space Θ. Denote the probability density function of $\mathbf{X}$ for a given value of θ as $f(\mathbf{x}|\theta)$ with $\mathbf{x} \in S$ and $\theta \in \Theta$.

In Bayesian analysis, named for the famous Thomas Bayes, one treats the parameter θ as a random variable, with a given probability density function $f(\theta)$ for $\theta \in \Theta$. The corresponding distribution is called the prior distribution of θ and is intended to reflect the knowledge of the parameter before gathering data. After observing $\mathbf{x} \in S$, one can use Bayes' theorem to compute the conditional probability density function of θ given $\mathbf{X} = \mathbf{x}$. The following is the derivation of Bayes' theorem.

The joint probability density function of $(\mathbf{X}, \theta)$ is the mapping on $S \times \Theta$ given as

$$(\mathbf{x}, \theta) \rightarrow f(\theta)f(\mathbf{x}|\theta).$$

The discrete probability density function f of $\mathbf{X}$ is given as

$$f(\mathbf{x}) = \sum_{\theta \in \Theta} f(\theta)f(\mathbf{x}|\theta), \quad \mathbf{x} \in S. \tag{7.52}$$

If the parameter has a continuous distribution, Eq. (7.52) is rewritten as

$$f(\mathbf{x}) = \int_{\Theta} f(\theta)f(\mathbf{x}|\theta), \quad \mathbf{x} \in S. \tag{7.53}$$

The conditional probability density function of θ is then

$$f(\theta|\mathbf{x}) = \frac{f(\theta)f(\mathbf{x}|\theta)}{f(\mathbf{x})}, \quad \theta \in \Theta, \ \mathbf{x} \in S. \tag{7.54}$$

The conditional distribution of θ given $\mathbf{x}$ is called the posterior distribution. Note that $f(\mathbf{x})$ is simply the normalizing constant for the function $f(\theta)f(\mathbf{x}|\theta)$. It may not be necessary to explicitly compute $f(\mathbf{x})$, if one can recognize the functional form of $f(\theta)f(\mathbf{x}|\theta)$ as a known distribution.

Assuming the data at the receiver is given as

$$\mathbf{x} = \mathbf{X}(w)\mathbf{g} + \mathbf{n}, \tag{7.55}$$

where w is the normalized CFO and $\mathbf{g}$ is the channel impulse response. $\mathbf{X}(w)$ contains the data information, and $\mathbf{n}$ is the white Gaussian noise with variance of σ^2.

Although the deterministic values of the CFO w and the channel $\mathbf{g}$ are not available, the statistical information about the CFO as well as channel may be known in many scenarios. For example, the channel $\mathbf{g}$ with L_g paths is usually modeled as a complex Gaussian random vector with the distribution given as

$$P(\mathbf{g}) = \frac{1}{\pi^{L_g}\det(\mathbf{Q})}e^{(-\mathbf{g}^H\mathbf{Q}^{-1}\mathbf{g})}, \qquad (7.56)$$

where $\mathbf{Q}$ is the channel covariance matrix containing the power delay profile information. The typical channel power delay profiles have been measured. Moreover, the maximum likelihood estimator is usually employed at the receiver, which can obtain the initial estimation of the CFO w. The distribution of w follows a Gaussian distribution given as [12]

$$P(w) = \frac{1}{\sqrt{2\pi}\sigma_w}e^{\left(-\frac{w^2}{2\sigma_w^2}\right)}, \qquad (7.57)$$

where σ_w^2 is the variance of the CFO distribution, which is related to the estimation accuracy.

According to Eq. (7.54), the posterior distribution of w is denoted as

$$P(w|\mathbf{x}) = \frac{P(\mathbf{x}|w)P(w)}{P(\mathbf{x})}. \qquad (7.58)$$

In order to obtain the estimation of w, one has to find the maximum $P(w|\mathbf{x})$. Since $P(\mathbf{x})$ is a constant, it only needs to find the maximum $P(\mathbf{x}|w)P(w)$.

When the noise variance σ^2 is known, the CFO posterior distribution $P(w|\mathbf{x})$ is

$$P(w|\mathbf{x},\sigma) \propto P(\mathbf{x}|w,\sigma)P(w) = \left(\int P(\mathbf{x}|w,\mathbf{g},\sigma)P(\mathbf{g})d\mathbf{g}\right)P(w), \quad (7.59)$$

where

$$P(\mathbf{x}|w,\mathbf{g},\sigma) = \frac{1}{(\pi\sigma^2)^K}\exp\left\{-\frac{[\mathbf{x}-\mathbf{X}(w)\mathbf{g}]^H[\mathbf{x}-\mathbf{X}(w)\mathbf{g}]}{\sigma^2}\right\}. \qquad (7.60)$$

According to Eqs. (7.56), (7.60), it can be obtained as

$$\int P(\mathbf{x}|w, \mathbf{g}, \sigma)P(\mathbf{g})d\mathbf{g} = \int \frac{1}{(\pi\sigma^2)^K}\exp\left\{-\frac{[\mathbf{x} - \mathbf{X}(w)\mathbf{g}]^H[\mathbf{x} - \mathbf{X}(w)\mathbf{g}]}{\sigma^2}\right\}$$
$$\times \frac{1}{\pi^{L_g}\det(\mathbf{Q})}e^{(-\mathbf{g}^H\mathbf{Q}^{-1}\mathbf{g})}. \tag{7.61}$$

Then,

$$\int P(\mathbf{x}|w, \mathbf{g}, \sigma)P(\mathbf{g})d\mathbf{g} \propto \frac{1}{(\pi\sigma^2)^K}\exp\left\{-\frac{\mathbf{x}^H\mathbf{C}(w)\mathbf{x}}{\sigma^2}\right\}$$
$$\times \int \exp\left\{-\frac{[\mathbf{g} - \mathbf{B}(w)\mathbf{x}]^H\mathbf{A}^{-1}[\mathbf{g} - \mathbf{B}(w)\mathbf{x}]}{\sigma^2}\right\}d\mathbf{g}, \tag{7.62}$$

where

$$\mathbf{C}(w) = \mathbf{I} - \mathbf{X}(w)[\mathbf{X}(w)^H\mathbf{X}(w) + \sigma^2\mathbf{Q}^{-1}]^{-1}\mathbf{X}(w)^H, \tag{7.63}$$
$$\mathbf{B}(w) = [\mathbf{X}(w)^H\mathbf{X}(w) + \sigma^2\mathbf{Q}^{-1}]^{-1}\mathbf{X}(w)^H, \tag{7.64}$$
$$\mathbf{A} = \mathbf{X}(w)^H\mathbf{X}(w) + \sigma^2\mathbf{Q}^{-1}, \tag{7.65}$$

Note that

$$\int \frac{1}{(\pi\sigma^2)^L\det(\mathbf{A})}\exp\left\{-\frac{[\mathbf{g} - \mathbf{B}(w)\mathbf{x}]^H\mathbf{A}^{-1}[\mathbf{g} - \mathbf{B}(w)\mathbf{x}]}{\sigma^2}\right\}d\mathbf{g} = 1, \tag{7.66}$$

and

$$\int P(\mathbf{x}|w, \mathbf{g}, \sigma)P(\mathbf{g})d\mathbf{g} \propto \frac{1}{(\pi\sigma^2)^{(K-L)}}\exp\left\{-\frac{\mathbf{x}^H\mathbf{C}(w)\mathbf{x}}{\sigma^2}\right\}. \tag{7.67}$$

Eq. (7.59) can then be rewritten as

$$P(w|\mathbf{x}, \sigma) \propto \exp\left\{-\frac{w^2}{2\sigma^2} - \frac{\mathbf{x}^H\mathbf{C}(w)\mathbf{x}}{\sigma^2}\right\}. \tag{7.68}$$

Hence, the estimation of w is obtained as

$$\hat{w} = \arg\min_w\left\{\frac{w^2}{2\sigma^2} + \frac{\mathbf{x}^H\mathbf{C}(w)\mathbf{x}}{\sigma^2}\right\}. \tag{7.69}$$

7.5 SUMMARY

As one of multicarrier systems, OQAM/FBMC suffers synchronization offset problem. Hence, reliable and accurate TO and CFO synchronization schemes are required for OQAM/FBMC systems. This chapter firstly analyzed the effects of TO and CFO in OQAM/FBMC systems. It was found that OQAM/FBMC is more robustness to synchronization offset compared with OFDM. Then, several traditional data–aided synchronization methods, such as TR1, TR2, LS (MLS), etc., were introduced in detail. Moreover, a new data–aided timing and CFO synchronization scheme was presented by exploiting the conjugate-symmetry property of a properly designed training sequence. Finally, we described a case of Bayesian CFO estimation as an example of the blind synchronization method.

REFERENCES

[1] Ciblat P, Serpedin E. A fine blind frequency offset estimator for OFDM/OQAM systems. IEEE Trans Signal Process 2004;52(1):291–6.
[2] Bolcskei H. Blind estimation of symbol timing and carrier frequency offset in wireless OFDM systems. IEEE Trans Commun 2001;49(6):988–99.
[3] Fusco T, Tanda M. Blind frequency-offset estimation for OFDM/OQAM systems. IEEE Trans Signal Process 2007;55(5):1828–38.
[4] Fusco T, Izzo L, Petrella A. Blind symbol timing estimation for OFDM/OQAM systems. IEEE Trans Signal Process 2009;57(12):4952–8.
[5] Fusco T, Petrella A, Tanda M. Data-aided symbol timing and CFO synchronization for filter bank multicarrier systems. IEEE Trans Wirel Commun 2009;8(5):2705–15.
[6] Tonello A, Rossi F. Synchronization and channel estimation for filtered multitone modulation. In: WPMC, Abano Terme; 2004.
[7] Saeedi-Sourck H, Sadri S, Yan W, Farhang-Boroujeny B. Near maximum likelihood synchronization for filter bank multicarrier systems. IEEE Wirel Commun Lett 2013;2(2):235–8.
[8] Farhang-Boroujeny B, Amini P. Packet format design and decision directed tracking methods for filter bank multicarrier systems. EURASIP J Adv Signal Process 2010;2010:307983.
[9] Farhang-Boroujeny B. Filter bank multicarrier modulation: a waveform candidate for 5G and beyond. In: Advances in electrical engineering; 2014.
[10] Hao L, Siohan P. Capacity analysis for indoor PLC using different multi-carrier modulation schemes. IEEE Trans Power Deliv 2009;25(1):113–24.
[11] Hao L, Gharba M, Siohan P. Impact of time and carrier frequency offsets on the OQAM-OFDM modulation scheme. Signal Process 2014;102:151–62.
[12] Cai K, Li X, Wu Y. Bayesian CFO estimation in OFDM systems. In: WCNC, Budapest; 2009.

CHAPTER 8

Channel Estimation

The channel estimation in offset quadrature amplitude modulation-based filter bank multicarrier (OQAM/FBMC) systems has a significant influence on the system performance [1]. However, due to the fact that the orthogonality condition of OQAM/FBMC systems only holds in the real field [2], the intrinsic imaginary interference would be induced to the transmitted real symbols at the receiver, which has to be considered for the channel estimation in OQAM/FBMC systems. Thus, existing orthogonal frequency division multiplexing (OFDM) channel estimation schemes cannot be directly employed in OQAM/FBMC systems.

Researchers have proposed various channel estimation schemes for OQAM/FBMC systems, which can be classified into two categories (i.e., frequency domain channel estimations and time domain channel estimations). Due to the simplicity, frequency domain channel estimations have attracted much attention. In [3–5], the scattered-based channel estimation methods were investigated for OQAM/FBMC systems. The preamble-based channel estimation scheme was proposed in [6] (i.e., interference approximation method [IAM]), which is similar to the frequency domain channel estimation in OFDM systems and simple single–tap equalizer can be employed to recover data symbols at the receiver. In [7–10], the imaginary-valued pilots were presented to improve the channel estimation. By relaxing the orthogonality constraint of OQAM/FBMC systems, the prototype filter was redesigned in [11] to improve the channel estimation. In [12], the researchers proposed an iterative channel estimation method to reduce the overhead of preamble pilots. In [13], a frequency-domain smoothing technique was proposed by minimizing the variance of the least square channel estimator. In [14], a review was presented to introduce the existing preamble structures and corresponding channel estimation schemes. However, frequency domain channel estimations would suffer performance degradation when the channel delay spread is large. To overcome this drawback, time domain channel estimations are proposed.

OQAM/FBMC for Future Wireless Communications
http://dx.doi.org/10.1016/B978-0-12-813557-0.00008-5

© 2018 Elsevier Ltd.
All rights reserved.

175

However, since the conventional OQAM/FBMC systems do not require an insertion of cyclic prefix (CP), most of the channel estimation methods are based on the assumption that the channel delay spread is very short to be able to employ the simple single-tap equalizer. Otherwise, complex multiple-tap equalizers have to be employed for channel equalization. To enable the simple single-tap equalizer, CP-OQAM/FBMC systems were proposed recently in [15]. So far, no related literatures focus on the channel estimation in CP-OQAM/FBMC systems. In this chapter, we will also present the channel estimation in CP-OQAM/FBMC systems.

The rest of this chapter is organized as follows. The preamble-based channel estimations in OQAM/FBMC are presented in Section 8.1. A scattered-based channel estimation in OQAM/FBMC is presented in Section 8.2. Finally, the channel estimations in CP-OQAM/FBMC are presented in Section 8.3, followed by summary in Section 8.4.

8.1 PREAMBLE-BASED CHANNEL ESTIMATION FOR OQAM/FBMC SYSTEMS

Preamble-based channel estimations in OQAM/FBMC systems can be classified into two categories (i.e., frequency domain channel estimations and time domain channel estimations). The main advantage of frequency domain channel estimations is simplicity. However, frequency domain channel estimations work well only when the channel delay spread is very small. When the channel delay spread is large, time domain channel estimations have to be employed in OQAM/FBMC systems. Accordingly, in this section, we will present the channel estimations in OQAM/FBMC systems.

8.1.1 Frequency Domain Channel Estimation

We consider an equivalent baseband system of OQAM/FBMC, in which the subcarrier number is K and subcarrier spacing is $1/T$, with T being the complex symbol interval. The transmitted symbol $a_k(n)$ is real-valued with frequency index k and time index n, and $T/2$ is the interval of real-valued symbols. $a_k(2m)$ and $a_k(2m+1)$ are obtained by taking the real and imaginary parts of a complex-valued symbol from 2^{2i}-quadrature amplitude modulation (QAM) constellation, respectively. $h(m)$ is a symmetrical real-valued pulse-shaping filter. Therefore, the equivalent baseband OQAM/FBMC signal is written as

$$s(m) = \sum_{k=0}^{K-1} \sum_{n \in \mathbb{Z}} a_k(n) \underbrace{h\left(m - n\frac{K}{2}\right) e^{j2\pi mk/K} e^{j\pi(k+n)/2}}_{h_{k,n}(m)}. \qquad (8.1)$$

Assuming a distortion-free channel, perfect reconstruction (PR) of real symbol $a_k(n)$ is obtained owing to the following orthogonality condition

$$\Re\left\{ \sum_{m=-\infty}^{\infty} h_{k,n}(m) h_{p,q}^*(m) \right\} = \delta_{k,p}\delta_{n,q}, \qquad (8.2)$$

where $\delta_{k,p} = 1$ if $k = p$; and $\delta_{k,p} = 0$ if $k \neq p$.

Let us denote

$$\zeta_{k,n}^{p,q} = \sum_{m=-\infty}^{\infty} g_{k,n}(m) g_{p,q}^*(m). \qquad (8.3)$$

When $(k, n) = (p, q)$, $\zeta_{k,n}^{p,q} = 1$, otherwise, $\zeta_{k,n}^{p,q}$ is an imaginary value.

The baseband version of the received signal could be written as

$$r(m) = g(m) * s(m) + \eta(m), \qquad (8.4)$$

where $g(m)$ is the multipath fading channel and $\eta(m)$ is the complex additive white Gaussian noise (AWGN) with zero mean and variance σ^2. The demodulation of the received signal at the (k, n)th index provides a complex symbol given as

$$\hat{a}_k(n) = \sum_{m=-\infty}^{\infty} r(m) h\left(m - n\frac{K}{2}\right) e^{-j2\pi mk/K} e^{-j\pi(k+n)/2}. \qquad (8.5)$$

Suppose that the symbol interval is much longer than the maximum channel delay spread L_g [6]. Therefore, the prototype filter function has relatively low variation over any time interval $[m, m + L_g]$, that is,

$$h(m) \approx h(m + \tau), \quad \text{for } \tau \in (0, L_g]. \qquad (8.6)$$

For simplicity, we assume that the preamble pilots have time index of 0 and they are not interfered by data symbols due to the insertion of enough zero symbols between the preamble pilots and data symbols. The received pilot on the kth subcarrier is then obtained as [6]

$$\hat{a}_k(0) \approx G_{k,0}\left(a_k(0) + a_k^c(0)\right) + \eta_k(0), \qquad (8.7)$$

where $a_k^c(0) = \sum_{i=0, i \neq k}^{K-1} \zeta_{i,0}^{k,0} a_i(0)$, $G_{k,0} = \sum_0^{L_g} g(k) e^{-2j\pi mk/K}$ represents the complex channel frequency response at the kth subcarrier, and $\eta_{k,n}$ is the demodulated noise at the (k, n)th position given as

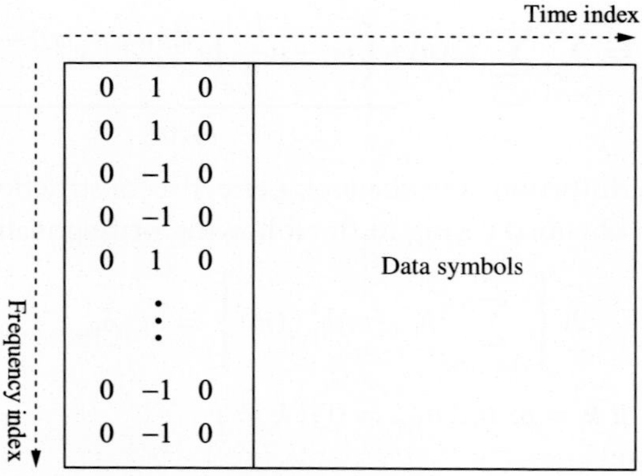

Time index

Fig. 8.1 The preamble structure for the IAM method.

$$\eta_k(n) = \sum_{m=-\infty}^{\infty} \eta(m)h\left(m - n\frac{K}{2}\right)e^{-j2\pi mk/K}e^{-j\pi(k+n)/2}. \quad (8.8)$$

With the frequency domain model given by Eq. (8.7), channel estimation based on the IAM method is given as [6]

$$\hat{G}_{k,0} = \frac{\hat{a}_k(0)}{a_k(0) + a_k^c(0)}. \quad (8.9)$$

Fig. 8.1 shows the preamble structure for the frequency domain channel estimations, which requires three columns of symbols for the pilot overhead.

8.1.2 Time Domain Channel Estimation

In this section, we present a preamble-based time domain model for channel estimation in the OQAM/FBMC systems. The key idea is to utilize frequency domain pilots to estimate the time domain channel impulse responses with no limitation on the length of the symbol interval.

By substituting Eqs. (8.1), (8.4) into Eq. (8.5), the demodulated symbol $\hat{a}_m(n)$ can be rewritten as

$$\hat{a}_k(n) = \sum_{l=0}^{L_g-1} \sum_{m=-\infty}^{\infty} \sum_{p=0}^{K-1}\sum_{q\in\mathbb{Z}} a_p(q)h\left(m - l - q\frac{K}{2}\right)h\left(m - n\frac{K}{2}\right)e^{j2\pi(p-k)m/K}$$
$$\times e^{j\pi(p+q-k-n)/2}e^{-j2\pi pl/K}g(l) + \eta_k(n). \quad (8.10)$$

The preamble pilots $\{a_k(0)\}_{k=0}^{K-1}$ are used for the channel estimation in the OQAM/FBMC system. In practice, the preamble pilots are interfered by the adjacent data symbols due to the overlapping structure of the adjacent symbols. Thus, zero symbols are usually inserted between the preamble pilots and data symbols to reduce the interference to an acceptable level. For simplicity, we ignore the intersymbol interference (ISI) to preamble pilots by inserting enough zero symbols between the preamble pilots and data symbols. The received pilot is then obtained as

$$\hat{a}_k(0) = \sum_{l=0}^{L_g-1} \sum_{m=-\infty}^{\infty} \sum_{p=0}^{K-1} a_p(0)h(m-l)h(k)e^{j2\pi(p-k)m/K}e^{j\pi(p-k)/2}e^{-j2\pi pl/K}g(l)$$
$$+ \eta_k(0). \tag{8.11}$$

For convenience, we rewrite Eq. (8.11) in vector notation as

$$\mathbf{r}_0 = \tilde{\mathbf{A}}\mathbf{g} + \boldsymbol{\eta}_0, \tag{8.12}$$

where $\mathbf{r}_0 = \begin{bmatrix} \hat{a}_0(0) & \hat{a}_1(0) & \cdots & \hat{a}_{K-1}(0) \end{bmatrix}^T$. $\mathbf{g} = \begin{bmatrix} g(0) & g(1) & \cdots & g(L_g-1) \end{bmatrix}^T$ is the vector of sampled channel response, and $\boldsymbol{\eta}_0 = [\eta_0(0) \ \eta_1(0) \ \cdots \ \eta_{K-1}(0)]^T$ is the noise vector. $\tilde{\mathbf{A}}$ is a $K \times L_g$ matrix determined by the preamble pilots and pulse-shaping filter, and the element on the kth row and lth column is given as

$$\tilde{\mathbf{A}}_{k,l} = \sum_{m=-\infty}^{\infty} \sum_{p=0}^{K-1} a_p(0)h(m-l)h(m)e^{j2\pi(p-k)m/K}e^{j\pi(p-k)/2} \times e^{-j2\pi pl/K}.$$
$$\tag{8.13}$$

In the channel estimation model in Eq. (8.12), the covariance matrix of the noise vector could be obtained as

$$\mathbf{V} = \begin{pmatrix} \sigma^2 & \sigma^2\zeta_{0,0}^{1,0} & \cdots & \sigma^2\zeta_{0,0}^{K-2,0} & \sigma^2\zeta_{0,0}^{K-1,0} \\ \sigma^2\zeta_{1,0}^{0,0} & \sigma^2 & \cdots & \sigma^2\zeta_{1,0}^{K-2,0} & \sigma^2\zeta_{1,0}^{K-1,0} \\ \vdots & \vdots & \ddots & \vdots & \vdots \\ \sigma^2\zeta_{K-2,0}^{0,0} & \sigma^2\zeta_{K-2,0}^{1,0} & \cdots & \sigma^2 & \sigma^2\zeta_{K-2,0}^{K-1,0} \\ \sigma^2\zeta_{K-1,0}^{0,0} & \sigma^2\zeta_{K-1,0}^{1,0} & \cdots & \sigma^2\zeta_{K-1,0}^{K-2,0} & \sigma^2 \end{pmatrix}. \tag{8.14}$$

The time domain channel estimation can be obtained by minimizing the weighted least square (WLS) error,

$$\text{WLSE} = \mathrm{E}\left[(\mathbf{r}_0 - \tilde{\mathbf{A}}\mathbf{g})^H \mathbf{V}^{-1}(\mathbf{r}_0 - \tilde{\mathbf{A}}\mathbf{g})\right], \tag{8.15}$$

where $\mathbf{V}^{-1}$ is the weighting matrix. The WLS could then be written as

$$\hat{\mathbf{g}}_{WLS} = \left(\tilde{\mathbf{A}}^H \mathbf{V}^{-1} \tilde{\mathbf{A}}\right)^{-1} \tilde{\mathbf{A}}^H \mathbf{V}^{-1} \mathbf{r}_0. \qquad (8.16)$$

Since both $\tilde{\mathbf{A}}$ and $\mathbf{V}$ are known for the receiver and dimension of $(\tilde{\mathbf{A}}^H \mathbf{V}^{-1} \tilde{\mathbf{A}})^{-1} \tilde{\mathbf{A}}^H \mathbf{V}^{-1}$ is $L_g \times K$, obviously the complexity of the WLS estimator is $\mathcal{O}(KL_g)$.

8.1.3 Simulation Results

Some simulations have been conducted in the OQAM/FBMC systems, where 128 subcarriers and 4-QAM modulation are employed. The pulse-shaping filters are extended Gaussian function (EGF) filters with the duration of $4T$:

$$g(t) = \begin{cases} z_{\varepsilon,\upsilon_0,\tau_0}(t), & -2T \leq t \leq 2T, \\ 0, & \text{else}, \end{cases} \qquad (8.17)$$

where $z_{\varepsilon,\upsilon_0,\tau_0}(t)$ could be obtained with [16],

$$z_{\varepsilon,\upsilon_0,\tau_0}(t) = \frac{1}{2} \left\{ \sum_{k=0}^{\infty} d_{k,\varepsilon,\upsilon_0} \left[G_{\varepsilon}\left(t + \frac{k}{\upsilon_0}\right) + G_{\varepsilon}\left(t - \frac{k}{\upsilon_0}\right) \right] \right\}$$
$$\times \sum_{l=0}^{\infty} d_{l,1/\varepsilon,\tau_0} \cos\left(2\pi l \frac{t}{\tau_0}\right), \qquad (8.18)$$

where $\tau_0 \upsilon_0 = \frac{1}{2}$, $0.528\upsilon_0^2 \leq \varepsilon \leq 7.568\upsilon_0$, and $d_{k,\varepsilon,\upsilon_0}$ is a real-valued coefficient and could be computed via the rules described in [17]. $G_{\varepsilon}(t)$ is the Gaussian function as

$$G_{\varepsilon}(t) = (2\varepsilon)^{1/4} e^{-\pi \varepsilon t^2}. \qquad (8.19)$$

A special case of EGF filter, $z_{1,1/\sqrt{2},1/\sqrt{2}}$, is called isotropic orthogonal transform algorithm (IOTA) function. As shown in Fig. 8.2, for the frame structure in the OQAM/FBMC systems, $\{a_{k,0}\}_{m=0}^{K-1}$ are the preamble pilots, which are randomly 1 or -1. The real-valued data symbol $a_{k,n}$ is obtained by extracting the real or imaginary part of the corresponding complex-valued symbol of 4-QAM constellation. To reduce the ISI to preamble pilots caused by data symbols, Z columns of zeros are inserted between the preamble pilots and data symbols. The interference to preamble pilots from data symbols is smaller when Z is larger.

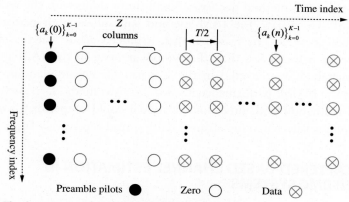

Fig. 8.2 The frame structure of the OQAM/FBMC system.

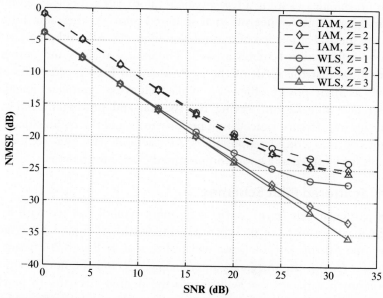

Fig. 8.3 NMSE performance comparison between the WLS and IAM method, with the IOTA filter.

Fig. 8.3 shows the NMSE performance comparison of the time and frequency domain channel estimations, where the IOTA filter is employed and $Z = 1, 2, 3$, respectively. Both the WLS and IAM methods exhibit performance floor at high signal-to-noise ratio when $Z = 1$, mainly due to the ISI to preamble pilots caused by data symbols. As Z increases, the performance floor of the WLS method decreases significantly along with the reduction of the ISI. Contrarily, the IAM method still suffers high performance floor when $Z = 3$. The reason is that, it indeed needs that the channel delay spread is much less than the symbol interval in the IAM method, Unfortunately, it is not valid in the scenario of our simulations, which results in the NMSE performance loss for the IAM method. In a word, the channel estimation with time domain model could outperform the IAM method, especially when the channel delay spread is relatively large.

8.2 SCATTERED-BASED CHANNEL ESTIMATION IN OQAM/FBMC SYSTEMS

In this section, we present a scattered pilot method for the channel estimation. We refer to this as coded auxiliary pilot (CAP) method. It builds on the idea of using a few auxiliary pilot (AP) symbols to eliminate the imaginary interference on each scattered pilot. At the same time, to keep the spectral efficiency of transmission high, the AP symbols are coded to carry information symbols as well. Theoretical analysis and numerical results show that the CAP method can maintain high spectral efficiency as well as high power efficiency, while achieving complete imaginary interference cancellation. Taking the spectral and power efficiency, the imaginary interference cancellation performance, and the coding/decoding complexity into consideration, the CAP method with two to four AP symbols is identified as a good compromised choice for practice.

8.2.1 Scattered Pilot Schemes

For a particular pair (m, n), $a_m(n)$ is the pilot symbol that we wish to use to obtain an estimate of the channel gain $G_m(n)$. If Eq. (8.20) is going to be used for this purpose, one will realize that in Eq. (8.20) beside the unknown $G_m(n)$, the imaginary interference $a_{m,n}^{(\star)}$ and the noise term $\eta_{m,n}$ are unknown variables that will disturb the desired estimate. In practice, nothing can be done to compensate for $\eta_{m,n}$, but $a_{m,n}^{(\star)}$ can be reduced, or even forced to zero, by cleverly imposing some relationship among the symbols around $a_m(n)$ that contribute to $a_{m,n}^{(\star)}$.

$$\tilde{a}_m(n) = G_m(n)(a_m(n) + ja_{m,n}^{(\star)}) + \eta_{m,n}. \tag{8.20}$$

The AP approach that has been proposed in [3], uses an auxiliary symbol adjacent to the pilot symbol $a_m(n)$ and it is selected such that its contribution to $a_{m,n}^{(\star)}$ will be the sign reverse of the sum of contributions from the reset of the interfering data symbols. This method, as noted earlier, is simple and can force $a_{m,n}^{(*)}$ to zero. However, as the numerical examples presented later show, it has a poor power efficiency.

The coding method that is proposed in [4], on the other hand, works as follows. A set of $N - 1$ PAM symbols, to be transmitted, are taken and mapped to N symbols that are placed at the positions around the pilot symbol $a_m(n)$. Using $\mathbf{d} = [d_1 \ d_2 \ \cdots \ d_{N-1}]^T$ to represent the vector containing these PAM symbols, and $\mathbf{s} = [s_1 \ s_2 \ \cdots \ s_N]^T$ to indicate the result after mapping, $\mathbf{d}$ and $\mathbf{s}$ are related according to the following equation.

$$\mathbf{s} = \mathbf{Cd}. \tag{8.21}$$

This may be thought as a linear coding/transformation step. Note that $\mathbf{C}$ is a matrix of size $N \times (N - 1)$. In other words it can be constructed as $\mathbf{C} = [\mathbf{c}_1 \ \mathbf{c}_2 \ \cdots \ \mathbf{c}_{N-1}]$, where $\mathbf{c}_1$ through $\mathbf{c}_{N-1}$ are a set of column vectors of size N.

In [4], $\mathbf{C}$ has been called coding matrix, because of obvious reasons. Also, it has been noted that by choosing $\mathbf{C}$ such that $\mathbf{C}^T\mathbf{C} = \mathbf{I}$, where $\mathbf{I}$ is the identity matrix, (i) the decoding matrix, to be used at the receiver, has the trivial form of $\mathbf{C}^T$; (ii) the coded symbols s_k hold the same power as the uncoded symbols d_k, thus, the imaginary interference reduction is achieved with an optimum power efficiency; and (iii) the decoding does not incur any noise enhancement.

Another observation that has been made in [4] is the following. The imaginary interference that is imposed by the elements of the coded vector $\mathbf{s}$ to the pilot symbol $a_m(n)$ can be calculated as

$$\theta = \boldsymbol{\gamma}^T\mathbf{s}, \tag{8.22}$$

where $\boldsymbol{\gamma} = [\gamma_1, \gamma_2, \ldots, \gamma_N]^T$ and γ_1 through γ_N are the imaginary interference coefficients, that is, the terms $\Im\left\{\zeta_{m+p,n+q}^{m,n}\right\}$, associated with s_1 through s_N, respectively. Since the goal of coding, here, is to force the imaginary interference θ to zero, Eq. (8.22) implies that

$$\boldsymbol{\gamma}^T\mathbf{C} = \mathbf{0}, \tag{8.23}$$

where $\mathbf{0}$ is a zero vector of compatible size to the left side.

Eq. (8.23) implies that the vector $\boldsymbol{\gamma}$ should be orthogonal with the vectors $\mathbf{c}_1$ through $\mathbf{c}_{N-1}$, that is, the columns of $\mathbf{C}$. This combined with the condition $\mathbf{C}^T\mathbf{C} = \mathbf{I}$, which was mentioned earlier, implies that $\mathbf{c}_1$ through $\mathbf{c}_{N-1}$ can be any set of orthogonal basis within the subspace orthogonal to $\boldsymbol{\gamma}$. In [4], the degrees of freedom in the selection $\mathbf{C}$ were made use of to develop a procedure for finding a coding matrix $\mathbf{C}$ that was sparse, hence, lend itself to a low complexity implementation.

It is important to note that in the coding method of [4] unless N, the dimension of $\mathbf{C}$, is made relatively large, the residual imaginary interference may remain significant, hence, lead to a poor channel estimation and accordingly a poor performance of the receiver. For instance, for the PHYDYAS and IOTA filters that we use in Section 8.2.5 to give our numerical results, one needs to set $N = 22$ and $N = 28$, respectively, before the residual imaginary interference could be reduced to below -40 dB of the pilot symbol energy. Complete cancellation of the imaginary interference will be only possible if N is set equal to 38, for PHYDYAS, and equal to 40 for IOTA.

For our reference here as well as in the rest of this chapter, Figs. 8.4 and 8.5 display the set of coded/AP/uncoded symbols that are surrounding a pilot symbol and may contribute to the imaginary interference, for PHYDYAS and IOTA cases, respectively. They are indexed for future references. Tables 8.1 and 8.2 list the respective imaginary interference coefficients, γ_k, of both. The numeric subscripts on γ_k's match the indexed symbols in Figs. 8.4 and 8.5.

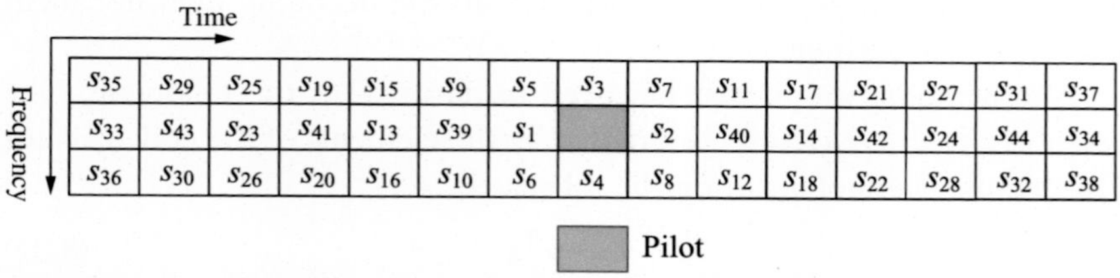

Time →

Frequency ↓

S_{35}	S_{29}	S_{25}	S_{19}	S_{15}	S_9	S_5	S_3	S_7	S_{11}	S_{17}	S_{21}	S_{27}	S_{31}	S_{37}
S_{33}	S_{43}	S_{23}	S_{41}	S_{13}	S_{39}	S_1		S_2	S_{40}	S_{14}	S_{42}	S_{24}	S_{44}	S_{34}
S_{36}	S_{30}	S_{26}	S_{20}	S_{16}	S_{10}	S_6	S_4	S_8	S_{12}	S_{18}	S_{22}	S_{28}	S_{32}	S_{38}

□ Pilot

Fig. 8.4 Indexed symbols surrounding a pilot for the PHYDYAS filter.

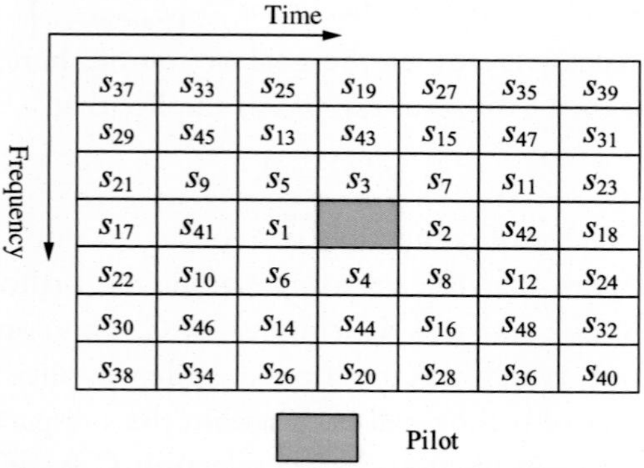

Time →

Frequency ↓

S_{37}	S_{33}	S_{25}	S_{19}	S_{27}	S_{35}	S_{39}
S_{29}	S_{45}	S_{13}	S_{43}	S_{15}	S_{47}	S_{31}
S_{21}	S_9	S_5	S_3	S_7	S_{11}	S_{23}
S_{17}	S_{41}	S_1		S_2	S_{42}	S_{18}
S_{22}	S_{10}	S_6	S_4	S_8	S_{12}	S_{24}
S_{30}	S_{46}	S_{14}	S_{44}	S_{16}	S_{48}	S_{32}
S_{38}	S_{34}	S_{26}	S_{20}	S_{28}	S_{36}	S_{40}

□ Pilot

Fig. 8.5 Indexed symbols surrounding a pilot for the IOTA filter.

Table 8.1 The imaginary interference coefficients when PHYDYAS filter is used

γ_1	-5.644×10^{-1}	γ_{12}	-1.250×10^{-1}	γ_{23}	-2.300×10^{-3}	γ_{34}	-6.735×10^{-6}
γ_2	5.644×10^{-1}	γ_{13}	-6.675×10^{-2}	γ_{24}	2.300×10^{-3}	γ_{35}	6.204×10^{-6}
γ_3	2.393×10^{-1}	γ_{14}	6.675×10^{-2}	γ_{25}	1.312×10^{-3}	γ_{36}	6.204×10^{-6}
γ_4	-2.393×10^{-1}	γ_{15}	4.288×10^{-2}	γ_{26}	1.312×10^{-3}	γ_{37}	6.204×10^{-6}
γ_5	2.058×10^{-1}	γ_{16}	4.288×10^{-2}	γ_{27}	1.312×10^{-3}	γ_{38}	6.204×10^{-6}
γ_6	2.058×10^{-1}	γ_{17}	4.288×10^{-2}	γ_{28}	1.312×10^{-3}	γ_{39}	0
γ_7	2.058×10^{-1}	γ_{18}	4.288×10^{-2}	γ_{29}	2.627×10^{-5}	γ_{40}	0
γ_8	2.058×10^{-1}	γ_{19}	5.362×10^{-3}	γ_{30}	-2.627×10^{-5}	γ_{41}	0
γ_9	1.250×10^{-1}	γ_{20}	-5.362×10^{-3}	γ_{31}	-2.627×10^{-5}	γ_{42}	0
γ_{10}	-1.250×10^{-1}	γ_{21}	5.362×10^{-3}	γ_{32}	-2.627×10^{-5}	γ_{43}	0
γ_{11}	1.250×10^{-1}	γ_{22}	-5.362×10^{-3}	γ_{33}	6.375×10^{-6}	γ_{44}	0

Table 8.2 The imaginary interference coefficients when IOTA filter is used

γ_1	-4.411×10^{-1}	γ_{13}	-3.805×10^{-2}	γ_{25}	1.027×10^{-2}	γ_{37}	4.633×10^{-4}
γ_2	4.411×10^{-1}	γ_{14}	-3.805×10^{-2}	γ_{26}	1.027×10^{-2}	γ_{38}	4.633×10^{-4}
γ_3	4.411×10^{-1}	γ_{15}	3.805×10^{-2}	γ_{27}	1.027×10^{-2}	γ_{39}	4.633×10^{-4}
γ_4	-4.411×10^{-1}	γ_{16}	3.805×10^{-2}	γ_{28}	1.027×10^{-2}	γ_{40}	4.633×10^{-4}
γ_5	2.280×10^{-1}	γ_{17}	-1.824×10^{-2}	γ_{29}	-1.573×10^{-3}	γ_{41}	0
γ_6	2.280×10^{-1}	γ_{18}	1.824×10^{-2}	γ_{30}	-1.573×10^{-3}	γ_{42}	0
γ_7	2.280×10^{-1}	γ_{19}	1.824×10^{-2}	γ_{31}	1.573×10^{-3}	γ_{43}	0
γ_8	2.280×10^{-1}	γ_{20}	-1.824×10^{-2}	γ_{32}	1.573×10^{-3}	γ_{44}	0
γ_9	3.805×10^{-2}	γ_{21}	1.027×10^{-2}	γ_{33}	1.573×10^{-3}	γ_{45}	0
γ_{10}	-3.805×10^{-2}	γ_{22}	1.027×10^{-2}	γ_{34}	-1.573×10^{-3}	γ_{46}	0
γ_{11}	3.805×10^{-2}	γ_{23}	1.027×10^{-2}	γ_{35}	1.573×10^{-3}	γ_{47}	0
γ_{12}	-3.805×10^{-2}	γ_{24}	1.027×10^{-2}	γ_{36}	-1.573×10^{-3}	γ_{48}	0

We note that increasing N brings with itself two drawbacks. First, the complexity of implementation of the coder and decoder increases with N^2. Second, when channel variation with time is fast and the scattered pilots need to be placed at a closer spacing, the coded symbols may contribute to the imaginary interference of the other nearby pilot symbols. In such cases, the simple linear coding Eq. (8.21) will no longer hold. Hence, more complex code designs and the associated coding and decoding steps need to be developed.

The CAP method that is introduced in Section 8.2.3 resolves both of the previous problems of the coding method, by applying a modification to the coded vector **s** which will remove the residual imaginary interference that is introduced by the surrounding data symbols which are not part of **s**. This modification, as our numerical results in Section 8.2.5 show, leads to a very significant improvement in the coding performance. At a minor cost in power inefficiency, we are able to remove the imaginary interference completely, even for very small values of N (in the range of 2–4).

Before ending this subsection a few comments should be made on the similarity and differences of the presentation of the coding matrix above and in [4]. We introduced the data vector $\mathbf{d} = [d_1 \ d_2 \ \cdots \ d_{N-1}]^T$, and accordingly the coding matrix **C** was chosen to be of size $N \times (N-1)$. The authors of [4], on the other hand, start with the modified data vector $\tilde{\mathbf{d}} = [d_1 \ d_2 \ \cdots \ d_{N-1} \ d_{N-1}]^T$, and accordingly the coding matrix **C** is chosen to be of size $N \times N$. Moreover, to satisfy Eq. (8.23), **C** has to be a rank $N-1$ matrix. This adds some complexity to the design of the coding matrix **C**. In addition, the repetition of d_{N-1} in $\tilde{\mathbf{d}}$ results in allocation of more power to d_{N-1} than the rest of the data symbols. The consequence of this change in power allocation is an additional burden on transmit power, without any obvious advantage on performance. Thus, the presentation of coding method, here, should be considered as a modification to that of [4].

8.2.2 Design of the Coding Matrix C

Lélé et al. [4] have discussed a procedure for selection of the coding matrix **C**, when the prototype filter is designed based on an IOTA [18]. In the case of IOTA design, the filter response is isotropic with respect to both time and frequency axes and as a result one finds that the imaginary interference coefficients γ_1 through γ_4 have the same amplitude. Other interference coefficients surrounding each pilot symbol also follow a similar isotropic symmetry. The procedure suggested in [4] takes advantage of this symmetry to find **C**. However, we note that IOTA designs are most appropriate for channels with fast variations. Other applications may have to consider other features of prototype filter and thus find other choices more appropriate. For instance, PHYDYAS group, who has extensively investigated FBMC for cellular applications, has found the filter design proposed by Mirabbasi and Martin [19] a better match [20, 21]. In the rest of this chapter, we refer to this choice of prototype filter as PHYDYAS.

For this design, as the numerical results in Table 8.1 show, the amplitude of γ_1 and γ_2 is different from that of γ_3 and γ_4. Hence, the design algorithm suggested in [4] for the selection of **C** is not applicable to PHYDYAS filter. Here, we present an alternative procedure for selection of **C** for the choices of N that are of interest to us. We discuss the choices of $N = 2$, 4, and 8. Other choices of N follows, similarly. Our procedure is also applicable to IOTA filter.

Next, we first proceed with the design of the coding matrix when the prototype filter is the PHYDYAS filter. The designs for the IOTA filter will be presented subsequently.

Coding Matrices for PHYDYAS Filter

For $N = 2$, we find from the numerical results in Table 8.1,

$$\gamma = \begin{bmatrix} \gamma_1 \\ \gamma_2 \end{bmatrix} = \begin{bmatrix} \gamma_1 \\ -\gamma_1 \end{bmatrix}. \tag{8.24}$$

Using Eq. (8.23), here, **C**, within a sign ambiguity, has the unique solution

$$\mathbf{C}_2 = \frac{1}{\sqrt{2}} \begin{bmatrix} 1 \\ 1 \end{bmatrix}. \tag{8.25}$$

For $N = 4$, we find from the numerical results in Table 8.1,

$$\gamma = \begin{bmatrix} \gamma_1 \\ -\gamma_1 \\ \gamma_3 \\ -\gamma_3 \end{bmatrix}. \tag{8.26}$$

Here, a trivial choice for the first two columns of **C** are

$$\mathbf{c}_1 = \frac{1}{\sqrt{2}} \begin{bmatrix} 1 \\ 1 \\ 0 \\ 0 \end{bmatrix} \quad \text{and} \quad \mathbf{c}_2 = \frac{1}{\sqrt{2}} \begin{bmatrix} 0 \\ 0 \\ 1 \\ 1 \end{bmatrix}.$$

Next, we note that c_3 should be a unit length vector orthogonal to $\boldsymbol{\gamma}$, c_1, and c_2. This within a sign ambiguity, has a unique solution. Finding this solution and combining that with c_1 and c_2, we get

$$\mathbf{C}_4 = \frac{1}{\sqrt{2}} \begin{bmatrix} 1 & 0 & 0.3903 \\ 1 & 0 & -0.3903 \\ 0 & 1 & 0.9207 \\ 0 & 1 & -0.9207 \end{bmatrix}. \tag{8.27}$$

For $N = 8$, we find from the numerical results in Table 8.2,

$$\boldsymbol{\gamma} = \begin{bmatrix} \gamma_1 \\ -\gamma_1\gamma_3 \\ -\gamma_3 \\ \gamma_5 \\ \gamma_5 \\ \gamma_5 \\ \gamma_5 \end{bmatrix}. \tag{8.28}$$

Following a procedure similar to the one that led to $\mathbf{C}_4$, but a bit more involved, we get

$$\mathbf{C}_8 = \frac{1}{\sqrt{2}} \begin{bmatrix} 1 & 0 & 0 & 0 & 0.3903 & 0 & 0.3948 \\ 1 & 0 & 0 & 0 & -0.3903 & 0 & -0.3948 \\ 0 & 1 & 0 & 0 & 0.9207 & 0 & -0.1674 \\ 0 & 1 & 0 & 0 & -0.9207 & 0 & 0.1674 \\ 0 & 0 & 1 & 0 & 0 & 0.7071 & 0.6388 \\ 0 & 0 & -1 & 0 & 0 & 0.7071 & 0.6388 \\ 0 & 0 & 0 & 1 & 0 & -0.7071 & 0.6388 \\ 0 & 0 & 0 & -1 & 0 & -0.7071 & 0.6388 \end{bmatrix}. \tag{8.29}$$

Here, the first four columns of $\mathbf{C}_8$ are selected to be orthogonal to $\boldsymbol{\gamma}$ and to each other, and, at the same time, each has minimum number of nonzero elements. The fifth column of $\mathbf{C}_8$, c_5, has been selected to have the minimum number of nonzero elements, and be orthogonal to $\boldsymbol{\gamma}$ and the first four columns of $\mathbf{C}_8$. Here, as may be noted from the result, to reduce the number of nonzero elements of c_5, we have forced its last four elements to zero and have found the unit length vector that is orthogonal to the vectors

$$\begin{bmatrix} 1 \\ 1 \\ 0 \\ 0 \end{bmatrix}, \quad \begin{bmatrix} 0 \\ 0 \\ 1 \\ 1 \end{bmatrix}, \quad \text{and} \quad \begin{bmatrix} 1 \\ -1 \\ 1 \\ -1 \end{bmatrix}.$$

Similar procedures have been applied to select the sixth and seventh columns of C_8.

Coding Matrices for IOTA Filter

Here, for $N = 2$, Eq. (8.24) is applicable, and accordingly C_2 is given by Eq. (8.25). For $N = 4$, Eq. (8.26) is simplified by replacing $\gamma_3 = \gamma_1$ according to Table 8.1. This leads to

$$C_4 = \frac{1}{\sqrt{2}} \begin{bmatrix} 1 & 0 & 0.7071 \\ 1 & 0 & -0.7071 \\ 0 & 1 & 0.7071 \\ 0 & 1 & -0.7071 \end{bmatrix}. \tag{8.30}$$

For $N = 8$, also, Eq. (8.28) is simplified by replacing $\gamma_3 = \gamma_1$. With this choice of $\boldsymbol{\gamma}$, a good design is

$$C_8 = \frac{1}{2} \begin{bmatrix} 1 & 1 & 0 & 0 & 1 & 0 & 0.4592 \\ 1 & 1 & 0 & 0 & -1 & 0 & -0.4592 \\ 1 & -1 & 0 & 0 & 1 & 0 & -0.4592 \\ 1 & -1 & 0 & 0 & -1 & 0 & 0.4592 \\ 0 & 0 & 1 & 1 & 0 & 1 & 0.8884 \\ 0 & 0 & -1 & -1 & 0 & 1 & 0.8884 \\ 0 & 0 & 1 & -1 & 0 & -1 & 0.8884 \\ 0 & 0 & -1 & 1 & 0 & -1 & 0.8884 \end{bmatrix}. \tag{8.31}$$

8.2.3 CAP Method

Let s_1 through s_L denote the data symbols (coded and uncoded) around the pilot symbol $a_{m,n}$ that contribute to the imaginary interference. As defined previously, the first N of these symbols are coded such that their combined imaginary interference be zero. We collected these coded symbols in the vector $\mathbf{s}$ that was defined earlier. We also used γ_1 through γ_N to denote the corresponding imaginary interference coefficients. Following the coding method of [4], $\mathbf{s}$ was selected such that $\boldsymbol{\gamma}^T\mathbf{s} = 0$. Note that this implies the vectors $\mathbf{s}$ and $\boldsymbol{\gamma}$ are orthogonal.

Next, we define the uncoded data symbol vector $\mathbf{s}' = [s_{N+1}\ s_{N+2}\ \cdots\ s_L]^T$ and the associated imaginary interference coefficient vector

$\boldsymbol{\gamma}' = [\gamma_{N+1}\gamma_{N+2} \cdots \gamma_L]^{\mathrm{T}}$. Accordingly, the imaginary interference that is introduced by the uncoded data symbols can be obtained as

$$\theta' = \boldsymbol{\gamma}'^{\mathrm{T}}\mathbf{s}'. \tag{8.32}$$

To compensate for and, hence, remove the imaginary interference θ', we add a modification vector $\boldsymbol{\delta}$ to $\mathbf{s}$. That is, the coded vector $\mathbf{s}$ of Eq. (8.21) is modified as

$$\mathbf{s} = \mathbf{Cd} + \boldsymbol{\delta}. \tag{8.33}$$

Next, we note that θ' will be removed, if $\boldsymbol{\delta}$ is chosen such that, for the modified $\mathbf{s}$ defined in Eq. (8.33),

$$\boldsymbol{\gamma}^{\mathrm{T}}\mathbf{s} = -\theta'. \tag{8.34}$$

Recalling Eq. (8.23), this simplifies to

$$\boldsymbol{\gamma}^{\mathrm{T}}\boldsymbol{\delta} = -\theta'. \tag{8.35}$$

At the same time, to pay a minimum power penalty for modification (Eq. 8.33), we choose $\boldsymbol{\delta}$ such that it has minimum norm. This implies $\boldsymbol{\delta}$ is selected by solving the following constrained minimization problem:

$$\min |\boldsymbol{\delta}|^2, \quad \text{subject to } \boldsymbol{\gamma}^{\mathrm{T}}\boldsymbol{\delta} = -\theta'. \tag{8.36}$$

This can be solved by using the method of Lagrange multipliers [22]. The result is

$$\boldsymbol{\delta} = -\frac{\theta'}{|\boldsymbol{\gamma}|^2}\boldsymbol{\gamma}. \tag{8.37}$$

We note that the construction in Eq. (8.33) builds based on two components. The first term on the right-hand side of Eq. (8.33) follows the method of [4] and codes the data symbols d_1 through d_{N-1} into a vector of length N such that the result incurs no imaginary interference to the pilot symbol $a_{m,n}$. The second term on the right-hand side of Eq. (8.33), on the other hand, follows the AP concept and $\boldsymbol{\delta}$ is selected such that the imaginary interference induced by the elements of $\mathbf{s}'$ is removed. It is in light of this observation that we call the method CAP.

8.2.4 Performance Analysis

In this section, the performance of the CAP method is analyzed and compared with the AP and coding methods in terms of spectral efficiency, power efficiency, residual imaginary interference, and pilot density.

Spectral Efficiency

For the AP method, one AP symbol is employed to eliminate the imaginary interference. Thus, the extra spectral overhead of the AP method for each pilot is one real-valued symbol. In other words, the channel estimate at each position of time-frequency space consumes two real-valued symbols, which is equivalent to one QAM (equivalently, one complex-valued) symbol. This, as was noted earlier, brings OQAM/FBMC in par with OFDM in terms of resources used for channel estimation.

For the coding method, $N - 1$ real-valued symbols are linearly coded into N symbols, which means, here also, one redundant symbol is used to remove the imaginary interference. Hence, the coding method also has the same spectral efficiency as the AP scheme and thus remains in par with OFDM. Obviously, the same is true for the CAP method that is presented in this chapter.

Power Efficiency

As a measure of power efficiency, we study the extra power that each of the three methods studied in this chapter uses to remove the imaginary interference to the pilot symbol. We refer to this extra power as *the power penalty*.

To derive the relevant equations for the power penalty of the three methods, we note that there is a total of $L - 1$ uncoded data symbols which are transmitted along with each pilot symbol and they contribute to the corresponding imaginary interference. We assume these data symbols are uncorrelated with one another and they are normalized to have a unit power. Hence, the total power required to transmit these data symbols is equal to $L - 1$.

In the coding method of [4], the $L - 1$ data symbols are divided into the pair of vectors $\mathbf{d}$ and $\mathbf{s}'$, that were defined in Section 8.2.3. The portion $\mathbf{s}'$ remains unchanged. Hence, does not contribute to any power penalty. The portion $\mathbf{d}$ is coded according to Eq. (8.21). The coded symbols s_1 through s_N will carry a total energy of $\mathbf{s}^T\mathbf{s} = \mathbf{d}^T\mathbf{C}^T\mathbf{C}\mathbf{d}$. Recalling the identity $\mathbf{C}^T\mathbf{C} = \mathbf{I}$, this implies that $\mathbf{s}^T\mathbf{s} = \mathbf{d}^T\mathbf{d}$, which in turn implies coding according to Eq. (8.21) does not introduce any power penalty. Hence, we conclude that

$$\varepsilon_{\text{coding}} = 0, \tag{8.38}$$

where $\varepsilon_{\text{coding}}$ is the power penalty of the coding method.

In the CAP scheme, similarly, the coding portion does not introduce any power penalty. However, the AP portion, that is, the addition of $\boldsymbol{\delta}$, is an power penalty. Accordingly,

$$\varepsilon_{CAP} = \mathbb{E}[\boldsymbol{\delta}^T \boldsymbol{\delta}]. \tag{8.39}$$

Substituting Eq. (8.37) in Eq. (8.39), we obtain

$$\varepsilon_{CAP} = \frac{1}{\boldsymbol{\gamma}^T \boldsymbol{\gamma}} \mathbb{E}[\theta'^2]. \tag{8.40}$$

Using Eq. (8.32) and recalling that we assume the elements of $\mathbf{s}'$ are uncorrelated and have unit power, it is straightforward to show that Eq. (8.40) simplifies to

$$\begin{aligned} \varepsilon_{CAP} &= \frac{\boldsymbol{\gamma}'^T \boldsymbol{\gamma}'}{\boldsymbol{\gamma}^T \boldsymbol{\gamma}} \\ &= \frac{\sum_{l=N+1}^{L} \gamma_l^2}{\sum_{l=1}^{N} \gamma_l^2}. \end{aligned} \tag{8.41}$$

The AP method may be thought to be a special case of CAP method, with $N = 1$. Hence, substituting $N = 1$ in Eq. (8.41), we obtain

$$\varepsilon_{AP} = \frac{\sum_{l=2}^{L} \gamma_l^2}{\gamma_1^2}. \tag{8.42}$$

Comparing Eqs. (8.38), (8.41), (8.42), one finds that

$$\varepsilon_{coding} \leq \varepsilon_{CAP} \leq \varepsilon_{AP}. \tag{8.43}$$

That is, in terms of power efficiency, the coding method has the best performance, and the AP method has the worst performance.

Residual Imaginary Interference

For a fixed pilot level, the quality of the channel estimate obtained through scattered pilots is influenced by how well the imaginary interference has been removed. Both AP and CAP methods remove the imaginary interference completely. Hence, using ξ to denote the power of the imaginary interference, we have

$$\xi_{AP} = \xi_{CAP} = 0. \tag{8.44}$$

In the coding method, any contribution from the data symbols s_{N+1} through s_L to the imaginary interference remain intact. Hence, considering the assumption that these symbols are uncorrelated and have unit power, we find that

$$\xi_{\text{coding}} = \sum_{l=N+1}^{L} \gamma_l^2. \tag{8.45}$$

We note that while both AP and CAP methods remove the imaginary interference perfectly, the coding method leaves some residual imaginary interference, hence, may result in a poor estimate of the channel and, accordingly, a degraded performance of the receiver.

Pilot Density

Sufficient pilots have to be inserted among data symbols to cope with the variation of channel along both time and frequency axes. Obviously, channels with faster variations demand for higher pilot density. Hence, the imaginary interference cancellation methods that allow higher density are more desirable. For the methods that were discussed in this chapter, as well as those in [3, 4], each set of auxiliary/coded symbols should only contribute to the imaginary interference of the respective pilot that they are attached to. Hence, the auxiliary/coded symbols attached to each pilot should be sufficiently spaced away from the rest of the pilots. This, in turn, implies the methods with smaller number of auxiliary/coded symbols offer a higher pilot density. Considering this point, one will conclude that AP method of [3] results in the highest pilot density, and the coding method of [4] leads to the lowest pilot density. This is further quantified, through specific examples, in the next section.

8.2.5 Numerical Results

The parameters that we use for the numerical calculations and simulations in this section are the following. The number of subcarriers in the OQAM/FBMC system is $K = 256$. Results are given for both the PHYDYAS filter and the IOTA filter. The multipath channel SUI-3 proposed by the IEEE802.16 broadband wireless access working group [23] is adopted for our simulations. Each data frame consists of nine OQAM symbols. The channel is randomly selected for each data frame, but kept fixed over the frame. Pilots are inserted at every four subcarriers, and at the fifth QAM symbol of the frame, as in Fig. 8.6. Moreover, the sum of power of each pilot symbol and the associated power penalty is twice the power of one complex data symbol. More comments on this choice is made below.

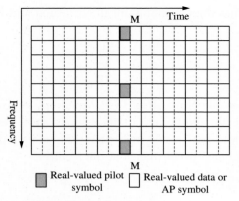

Fig. 8.6 Frame structure adopted in the simulations.

The channel estimates obtained at pilot positions are interpolated using linear interpolator to obtain the channel gain at the rest of the subcarrier. We examine three choices of the number of coded symbols: $N = 2$, 4, and 8. The coded symbols are selected following the indexing order in Figs. 8.4 and 8.5 (e.g., for $N = 4$), the coded symbols are s_1, s_2, s_3, and s_4.

The three methods that are studied in this chapter (AP, coding, and CAP) have the same spectral efficiency. All use one extra redundant real symbol to compensate for the imaginary interference.

A summary of our numerical results that indicate the values of power penalty (as a measure of power efficiency), and residual imaginary interference are presented in Tables 8.3 and 8.4 for the PHYDYAS and IOTA pulse shapes, respectively. As discussed in the previous section and numerically reflected in these tables, while the coding method incurs no

Table 8.3 Performance comparison of AP, coding, and CAP methods based on the PHYDYAS pulse shape

	AP [3]	Coding [4]			CAP, presented in this chapter		
Number of coded/AP symbols (N)	1	2	4	8	2	4	8
Power penalty	2.1388	0	0	0	0.5693	0.3302	0.0855
Residual imaginary interference power	0	0.3628	0.2483	0.0789	0	0	0

Table 8.4 Performance comparison of AP, coding, and CAP methods based on the IOTA pulse shape

	AP [3]				Coding [4]		CAP, presented in this chapter
Number of coded/AP symbols (N)	1	2	4	8	2	4	8
Power penalty	4.1392	0	0	0	1.5696	0.2848	0.0139
Residual imaginary interference power	0	0.6108	0.2216	0.0137	0	0	0

power penalty, the AP method has to put a significant amount of power into the AP to remove the imaginary interference. This power penalty reduces significantly by applying the CAP method. It is reduced by a factor of 3.7, 6.5, and 25 for the choices $N = 2$, 4, and 8, respectively, for the PHYDYAS filter. Similar, or better, improvements are also observed in the case of IOTA filter. The coding method, on the other hand, suffers from a significant degradation due to the residual (uncanceled) imaginary interference. The bit error rate (BER) results that are presented next provide more insight to the impact of these numbers on the performance of the three methods.

Table 8.5 presents a summary of the maximum pilot density that can be achieved for both cases of PHYDYAS and IOTA filters. The pilot density is measured as the ratio of the number of pilot symbols over the number of real–valued PAM data plus coded/AP and pilot symbols. The results are given for two choices of a parameter γ_{th}. This parameter is a threshold level that is used to decide when an imaginary interference coefficient γ_k is less than γ_{th} its contribution to imaginary interference is considered negligible,

Table 8.5 Pilot density for the PHYDYAS and IOTA pulse-shaping filters

	Pilot density			
N	$\gamma_{th} = 10^{-2}$		$\gamma_{th} = 10^{-3}$	
	PHYDYAS	IOTA	PHYDYAS	IOTA
1	$\frac{1}{5\times2} = 0.100$	$\frac{1}{5\times4} = 0.050$	$\frac{1}{7\times2} = 0.071$	$\frac{1}{5\times4} = 0.050$
2	$\frac{1}{5\times2} = 0.100$	$\frac{1}{5\times4} = 0.050$	$\frac{1}{7\times2} = 0.071$	$\frac{1}{5\times4} = 0.050$
4	$\frac{1}{5\times3} = 0.067$	$\frac{1}{5\times5} = 0.040$	$\frac{1}{7\times3} = 0.048$	$\frac{1}{5\times5} = 0.040$
8	$\frac{1}{5\times3} = 0.067$	$\frac{1}{5\times5} = 0.040$	$\frac{1}{7\times3} = 0.048$	$\frac{1}{5\times5} = 0.040$

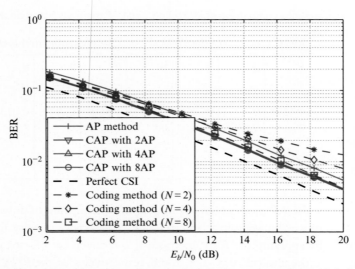

Fig. 8.7 BER performance when data symbols belong to a 4-QAM alphabet, with PHYDYAS filter.

hence, is set equal to zero in evaluating the pilot density. As expected the pilot density depends on γ_{th} and the number of coded/AP symbols (N). It decreases as N increases and/or as γ_{th} decreases.

Figs. 8.7–8.9 present the BER results that we have obtained for comparing the AP method of [3], the coding method of [4], and the CAP method that is developed in this chapter, and when the prototype filter is the PHYDYAS. The results have taken into account the power consumed for pilots and APs and accordingly have calculated the power per information bit over noise variance (E_b/N_0). In order to compare the three methods on the same basis, we used the same power for each pilot plus its associated AP. In the coding method, where there is no AP, all the power is given to the pilot. We have followed the recommended pilot power boosting method [24] and each pilot plus the associated AP (equivalently, the power penalty) have been allocated a power which is 3 dB higher than the power of each QAM data symbol. Moreover, as a base for comparing our results against, we have presented the BER curves that correspond to the case where channel state information is known perfectly. The results are based on sufficient runs so that each point on the BER curves is concluded after observing at least 10,000 bit errors.

As one would expect from the residual imaginary interference power values presented in Table 8.3, the coding method, in general, performs poorly. In the case of 4-QAM symbols, when the number of coded symbols

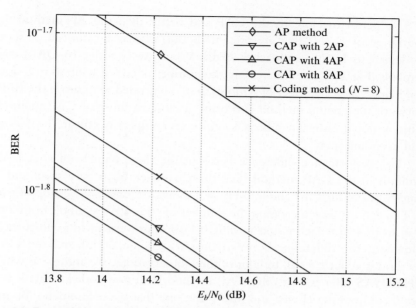

Fig. 8.8 Zoomed-in version of a portion of Fig. 8.7.

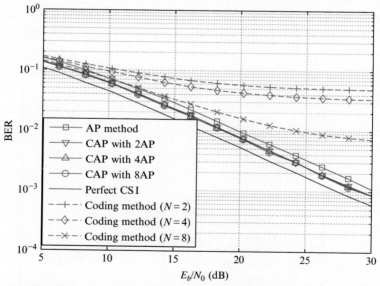

Fig. 8.9 BER performance when data symbols belong to a 16-QAM alphabet, with PHYDYAS filter.

is 8, it performs close to the CAP and better than the AP method for the range of E_b/N_0 that are presented. But, as may be noticed, it starts to deviate as E_b/N_0 approaches 25 dB. When we consider 16-QAM data symbols (Fig. 8.9), even increasing the number of coded symbols to 8 does not help. A larger number of symbols should be coded to improve the BER curve of the coding method and bring it close to those of CAP method. This, as noted earlier, will be at a significant complexity cost in the coding and decoding steps.

The performance difference between the three choices of coded/AP symbols in the CAP method (i.e., the parameter N) is very small and is hard to see in Figs. 8.7 and 8.9. Fig. 8.8 shows a zoomed-in version of part of Fig. 8.7. As seen increasing the number of coded/AP symbols from 2 to 4 or from 4 to 8 each results in only 0.1 dB improvement in performance. However, the difference between the AP scheme of [3] and the CAP method is about 1.5 dB. In light of these observations, we argue that when PHYDYAS filter is used, the CAP method with two coded symbols is a good compromise choice and thus is the one that we recommend for any practical system.

The BER results when IOTA is used as the prototype filter are presented in Figs. 8.10 and 8.11 for the symbol constellations 16-QAM and 64-QAM, respectively. Here, for the AP method, the power penalty is so high that it

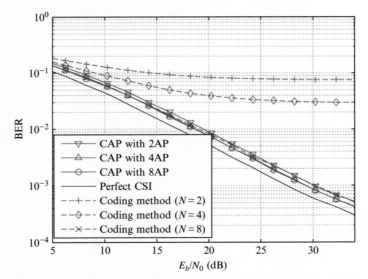

Fig. 8.10 BER performance when data symbols belong to a 16-QAM alphabet, with IOTA filter.

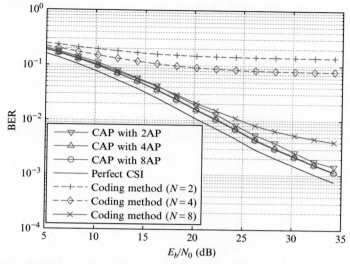

Fig. 8.11 BER performance when data symbols belong to a 64-QAM alphabet, with IOTA filter.

even exceeds the total power allocated for the pilot and AP, therefore the AP method is not a viable method under the current parameter setup. The performance of the CAP method with $N = 2$ is slightly degraded compared with that of the PHYDAYS filter, due to the fact that a larger power penalty incurs here. It is also observed that due to the isotropic distribution of imaginary interference over time-frequency grid, the performance of the coding method with $N = 8$ is better than that of the PHYDAS filter under 16-QAM modulation. This is in line with the lower imaginary interference power when the numerical results in Tables 8.3 and 8.4 are compared; 0.0789 against 0.0137. Here, the coding method, with $N = 8$, at BER $= 10^{-3}$ is comparable with the CAP method, with $N = 2$. However, if we move on to a larger constellation, like 64-QAM, the coding method experiences a significant degradation; see Fig. 8.11.

8.3 CHANNEL ESTIMATION FOR CP-OQAM/FBMC SYSTEMS

Similar to OFDM systems, CP-OQAM/FBMC systems convert a linear convolutional channel to a circular convolutional channel and therefore, make it possible to employ the simple single-tap equalizer for the channel equalization in the frequency domain. In this section, we will present the channel estimation in CP-OQAM/FBMC systems.

8.3.1 System Model

We consider an equivalent baseband version of the CP-OQAM/FBMC system with K subcarriers. The transmitted symbols are grouped into blocks with size $2N_d$. By the cyclic convolution between the transmitted symbols and filter, the transmitted signal of one block before the insertion of CP can be written as [15, 25]

$$s(n) = \sum_{k=0}^{K-1} \sum_{m=0}^{2N_d-1} j^{m+k} e^{j\pi \, mk} a_k(m) h_k(((n - mK/2))_{N_c}), \qquad (8.46)$$

where $N_c = KN_d$ and $0 \leq n \leq N_c - 1$, $((\cdot))_{N_c}$ denotes the mod N_c operation, and $a_k(m)$ are real transmitted symbols from the real and imaginary parts of complex symbols in a QAM constellation. The filter $h_k(n)$ on $(k + 1)$th subcarrier is

$$h_k(n) = \begin{cases} f_k(n), & 0 \leq n \leq \frac{\alpha K}{2}, \\ f_k(n - N_c + 1), & N_c - 1 - \frac{\alpha K}{2} \leq n < N_c - 1, \\ 0, & \text{else,} \end{cases} \qquad (8.47)$$

where

$$f_k(n) = h(n) e^{j2\pi \, kn/K}, \qquad -\frac{\alpha K}{2} \leq n \leq \frac{\alpha K}{2}, \qquad (8.48)$$

where $h(n)$ is a symmetrical real-valued pulse-shaping filter and satisfies the PR condition, whose length is $\alpha K + 1$, generally, α is set to 4. We rewrite Eq. (8.46) as

$$s(n) = \sum_{k=0}^{K-1} \sum_{m=0}^{2N_d-1} j^k e^{j\pi \, m(k+0.5)} a_k(m) h_k(((n - mK/2))_{N_c}), \qquad (8.49)$$

which will be used later for a compact vector-matrix representation.

In CP-OQAM/FBMC systems, CP can be utilized to avoid the ISI and intercarrier interference. The last N_g signal samples of $s(n)$ are inserted prior to the block before transmission, and it is required that $N_g \geq L - 1$. After inserting CP, the transmitted signal can then be obtained as

$$s_g(n) = \begin{cases} s(N_c - N_g + n), & 0 \leq n \leq N_g - 1, \\ s(n - N_g), & N_g \leq n \leq N_c + N_g - 1. \end{cases} \qquad (8.50)$$

Suppose a multipath fading channel $g(n), n = 0, 1, \ldots, L_g - 1$, interfered by the complex AWGN $\eta(n)$ with zero mean and variance σ^2. The received signal $r_g(n)$ can be obtained as

$$r_g(n) = s_g(n) \star g(n) + \eta(n), \tag{8.51}$$

where $\star$ stands for the linear convolution. After removing the CP, the received signal is written as

$$r(n) = r_g(n + N_g), \quad 0 \le n \le N_c - 1. \tag{8.52}$$

The demodulated symbol of received signal can then be obtained as

$$\hat{a}_k(m) = \sum_{n=0}^{N_c-1} (-j)^k e^{-j\pi m(k+0.5)} h_k(((mK/2 - n))_{N_c}) r(n). \tag{8.53}$$

For simplicity, we can rewrite Eq. (8.49) in matrix form

$$\mathbf{s} = \sum_{k=0}^{K-1} j^k \mathbf{H}_k \mathbf{M} \boldsymbol{\Sigma}_k \mathbf{a}_k, \tag{8.54}$$

where $\mathbf{a}_k = [a_k(0), a_k(1), \ldots, a_k(2N_d - 1)]^T$, $\mathbf{s} = [s(0), s(1), \ldots, s(N_c - 1)]^T$, $\boldsymbol{\Sigma}_k = \mathrm{diag}([1, e^{j\pi(k+0.5)}, \ldots, e^{j\pi(2N_d-1)(k+0.5)}])$, $\mathbf{M} = [\mathbf{e}_0, \mathbf{e}_{K/2}, \ldots, \mathbf{e}_{(2N_d-1)K/2}]$, $\mathbf{H}_k = \mathrm{circ}(\mathbf{h}_k)$ with $\mathbf{h}_k = [h_k(0), h_k(1), \ldots, h_k(N_c - 1)]^T$. Since $\mathbf{H}_k$ is a circular matrix, the following decomposition holds

$$\mathbf{H}_k = \mathbf{W}_{N_c}^H \boldsymbol{\Lambda}_k \mathbf{W}_{N_c}, \tag{8.55}$$

where $\mathbf{W}_{N_c}$ stands for the N_c-point discrete Fourier transform (DFT) matrix with the (m, n)th element $[\mathbf{W}_{N_c}]_{m,n} = \frac{1}{\sqrt{N_c}} e^{-j2\pi mn/N_c}$, and $\boldsymbol{\Lambda}_k = \mathrm{diag}(\boldsymbol{\lambda}_k)$ with $\boldsymbol{\lambda}_k = \sqrt{N_c} \mathbf{W}_{N_c} \mathbf{h}_k$. Eq. (8.54) can then be rewritten as

$$\mathbf{s} = \mathbf{W}_{N_c}^H \sum_{k=0}^{K-1} \mathbf{P}_k \mathbf{a}_k, \tag{8.56}$$

where $\mathbf{P}_k = j^k \boldsymbol{\Lambda}_k \mathbf{W}_{N_c} \mathbf{M} \boldsymbol{\Sigma}_k$. The following equation holds for the orthogonality of CP-OQAM/FBMC systems [15]

$$\Re\{\mathbf{P}_k^H \mathbf{P}_m\} = \begin{cases} \mathbf{I}_{2N_d}, & k = m, \\ \mathbf{0}_{2N_d}, & k \ne m, \end{cases} \tag{8.57}$$

where $\mathbf{0}_{2N_d}$ is zero matrix with $2N_d \times 2N_d$ dimension. After removing the CP, the received signal is

$$\mathbf{r} = \mathbf{G}_c\mathbf{s} + \boldsymbol{\eta}, \tag{8.58}$$

where $\mathbf{r} = [r(0), r(1), \dots, r(N_c - 1)]^T$, $\mathbf{G}_c = \text{circ}(\mathbf{g}_c)$ with $\mathbf{g}_c = [g(0), g(1), \dots, g(L_g - 1), 0, 0, \dots, 0]^T$ that is an $N_c \times 1$ vector, and $\boldsymbol{\eta} = [\eta(N_g), \eta(N_g + 1), \dots, \eta(N_g + N_c - 1)]^T$ is the channel noise. After N_c-point DFT operation on the received signal $\mathbf{r}$, it can be written as

$$\mathbf{r}_f = \mathbf{W}_{N_c}\mathbf{r} = \text{diag}(\mathbf{G})\sum_{k=0}^{K-1}\mathbf{P}_k\mathbf{a}_k + \mathbf{W}_{N_c}\boldsymbol{\eta}, \tag{8.59}$$

where $\mathbf{G} = [G(0), G(1), \dots, G(N_c - 1)]^T$ with $G(m) = \sum_{l=0}^{N_c-1} g(l)e^{-j2\pi ml/N_c}$, the channel frequency response. Obviously, since $\text{diag}(\mathbf{G})$ is a diagonal matrix, the simple single-tap equalizer can be employed when the channel impulse response is known at the receiver and the demodulated symbols are then

$$\hat{\mathbf{a}}_k = \mathbf{P}_k^H[\text{diag}(\mathbf{G})]^{-1}\mathbf{r}_f, \tag{8.60}$$

where $\hat{\mathbf{a}}_k = [\hat{a}_k(0), \hat{a}_k(1), \dots, \hat{a}_k(2N_d - 1)]^T$. Without the channel noise, it can be obtained

$$\Re\left\{\hat{\mathbf{a}}_k\right\} = \mathbf{a}_k. \tag{8.61}$$

Under the multipath fading channel, channel estimation and equalization are essential to recover the transmitted symbols perfectly.

8.3.2 WLS Channel Estimation in CP-OQAM/FBMC Systems

Fig. 8.12 depicts the WLS channel estimator and equalizer in CP-OQAM/FBMC systems. Firstly, a channel estimation model is presented on the

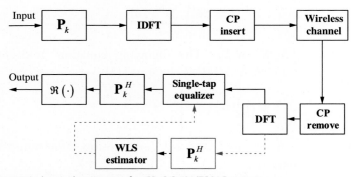

Fig. 8.12 WLS channel estimator for CP-OQAM/FBMC systems.

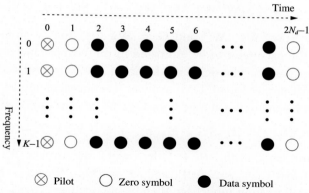

Fig. 8.13 One input block for the WLS estimator in CP-OQAM/FBMC systems.

channel frequency response by using transmitted pilot symbols. Then, by minimizing the WLSE, the WLS channel estimator can be obtained. Finally, with the estimated channel, a single-tap equalizer is employed to recover the data symbols.

As shown in Fig. 8.13, $a_k(0), k = 0, 1, \ldots, K - 1$, are the transmitted pilots whose values are randomly 1 or -1, which is similar to the block pilot in OFDM systems. Similar to OQAM/FBMC systems, due to the overlapping structure of the adjacent symbols, the demodulated pilot symbols in CP-OQAM/FBMC systems are interfered by the adjacent data symbols. To reduce the interference to an acceptable level, we can insert zero symbols between the preamble pilots and data symbols, that is, $a_k(1) = a_k(2N_d - 1) = 0, k = 0, 1, \ldots, K - 1$. Let $\mathbf{S} = \text{circ}(\mathbf{s})$. Since both $\mathbf{G}_c$ and $\mathbf{S}$ are circular matrices, Eq. (8.58) can be rewritten as

$$\mathbf{r} = \mathbf{G}_c \mathbf{s} + \boldsymbol{\eta} = \mathbf{S} \mathbf{g}_c + \boldsymbol{\eta} = \hat{\mathbf{S}} \mathbf{g} + \boldsymbol{\eta}, \tag{8.62}$$

where $\mathbf{g} = [g(0), g(1), \ldots, g(L_g - 1)]^T$ is the multipath channel and $\hat{\mathbf{S}}$ is the $N_c \times L_g$ submatrix of the $N_c \times N_c$ matrix $\mathbf{S}$ (i.e., the (m, n)th entry $[\hat{\mathbf{S}}]_{m,n} = [\mathbf{S}]_{m,n}$). Note that since $\hat{\mathbf{S}}$ is generated by both of the pilots and the unknown data symbols, it is unknown at the receiver. Of course, if we regard the whole block symbols as pilots, $\hat{\mathbf{S}}$ can be considered known; however, the pilot overhead will be extremely high, which is unacceptable. Furthermore, it is not necessary for $\hat{\mathbf{S}}$ to be known to solve for the channel as we shall see in the following section in more details. Thus, Eq. (8.62) cannot be applied to the channel estimation directly. Let us define an $N_c \times L_g$ matrix $\mathbf{W}$ with the (m, n)th entry $[\mathbf{W}]_{m,n} = e^{-j2\pi mn/N_c}$. Then, Eq. (8.62) can be rewritten as

$$\mathbf{r} = \hat{\mathbf{S}}(\mathbf{W}^H\mathbf{W})^{-1}\mathbf{W}^H\mathbf{W}\mathbf{g} + \boldsymbol{\eta} = \tilde{\mathbf{S}}\mathbf{G} + \boldsymbol{\eta}, \qquad (8.63)$$

where $\tilde{\mathbf{S}} = \hat{\mathbf{S}}(\mathbf{W}^H\mathbf{W})^{-1}\mathbf{W}^H$ and $\mathbf{G} = \mathbf{W}\mathbf{g}$ (i.e., the N_c-point DFT of channel) is the parameter to be estimated in this chapter and used for the channel equalization. The signal model in Eq. (8.63) is similar to the time domain signal model for OFDM systems. Its equivalent model similar to the frequency domain signal model for OFDM systems can be expressed as

$$\hat{\gamma}_k(0) = \boldsymbol{\Theta}_{k,0}\mathbf{r} = \boldsymbol{\Theta}_{k,0}\tilde{\mathbf{S}}\mathbf{G} + \boldsymbol{\Theta}_{k,0}\boldsymbol{\eta}, \quad k = 0, 1, \dots, K - 1, \qquad (8.64)$$

where vector $\boldsymbol{\Theta}_{k,p}$ is the $(p + 1)$th row of the matrix $\mathbf{P}_k^H\mathbf{W}_{N_c}$. Let $\mathbf{y} = [\hat{\gamma}_0(0), \hat{\gamma}_1(0), \dots, \hat{\gamma}_{K-1}(0)]^T$ and $\boldsymbol{\Theta} = [\boldsymbol{\Theta}_{0,0}^T, \boldsymbol{\Theta}_{1,0}^T, \dots, \boldsymbol{\Theta}_{K-1,0}^T]^T$. The channel estimation model can then be obtained as

$$\mathbf{y} = \boldsymbol{\Theta}\tilde{\mathbf{S}}\mathbf{G} + \boldsymbol{\Theta}\boldsymbol{\eta}. \qquad (8.65)$$

Different from a CP-based OFDM system, in the following, we show that although matrix $\boldsymbol{\Theta}\tilde{\mathbf{S}}$ may not be diagonal, it can be approximately considered known when $a_k(0)$, $k = 0, 1, \dots, K - 1$, are known as pilots and $a_k(1) = a_k(2N_d - 1) = 0$, $k = 0, 1, \dots, K - 1$. For the $(k + 1)$th row of $\boldsymbol{\Theta}\tilde{\mathbf{S}}$, it can be written as

$$\boldsymbol{\Theta}_{k,0}\tilde{\mathbf{S}} = \boldsymbol{\Theta}_{k,0}\hat{\mathbf{S}}(\mathbf{W}^H\mathbf{W})^{-1}\mathbf{W}^H$$

$$= \mathbf{s}^T\left[\tilde{\boldsymbol{\Theta}}_{k,0,0}^T, \tilde{\boldsymbol{\Theta}}_{k,0,1}^T, \dots, \tilde{\boldsymbol{\Theta}}_{k,0,L-1}^T\right](\mathbf{W}^H\mathbf{W})^{-1}\mathbf{W}^H, \qquad (8.66)$$

where $\tilde{\boldsymbol{\Theta}}_{k,0,p}$ is a row vector with $\tilde{\boldsymbol{\Theta}}_{k,0,p}(m) = \boldsymbol{\Theta}_{k,0}(((m + p))_{N_c})$. For $\mathbf{s}^T\tilde{\boldsymbol{\Theta}}_{k,0,p}^T$, it can be written as

$$\mathbf{s}^T\tilde{\boldsymbol{\Theta}}_{k,0,p}^T = \sum_{l=0}^{K-1} \mathbf{a}_l^T\mathbf{P}_p^T\mathbf{W}_{N_c}^*\tilde{\boldsymbol{\Theta}}_{k,0,p}^T$$

$$= \sum_{l=0}^{K-1} \mathbf{a}_l^T\left[\boldsymbol{\Theta}_{l,0}^H, \boldsymbol{\Theta}_{l,1}^H, \dots, \boldsymbol{\Theta}_{l,2N_d-1}^H\right]^T\tilde{\boldsymbol{\Theta}}_{k,0,p}^T$$

$$= \sum_{l=0}^{K-1} \boldsymbol{\Theta}_{l,0}^*\tilde{\boldsymbol{\Theta}}_{k,0,p}^Ta_l(0) + \sum_{l=0}^{K-1} \boldsymbol{\Theta}_{l,1}^*\tilde{\boldsymbol{\Theta}}_{k,0,p}^Ta_l(1)$$

$$+ \sum_{l=0}^{K-1} \boldsymbol{\Theta}_{l,2N_d-2}^*\tilde{\boldsymbol{\Theta}}_{k,0,p}^Ta_l(2N_d - 1)$$

$$+ \sum_{l=0}^{K-1}\sum_{m=2}^{2N_d-2} \boldsymbol{\Theta}_{l,m}^*\tilde{\boldsymbol{\Theta}}_{k,0,p}^Ta_l(m). \qquad (8.67)$$

Table 8.6 Value of $\Theta_{l,m}^* \tilde{\Theta}_{k,n,0}^T$, with IOTA filter

$(l-k,$ $m-n)$	-2	-1	0	1	2
-2	$-5.1761e-07$	$0.0380j$	$-1.4875e-08$	$-0.0380j$	$-5.1761e-07$
-1	$-0.0380j$	$-0.2280j$	$-0.4411j$	$-0.2280j$	$-0.0380j$
0	$-1.4408e-05$	$0.4411j$	1	$-0.4411j$	$-1.4408e-05$
1	$0.0380j$	$-0.2280j$	$0.4411j$	$-0.2280j$	$0.0380j$
2	$-5.1761e-07$	$0.0380j$	$-1.4875e-08$	$-0.0380j$	$-5.1761e-07$

$\Theta_{l,m}^* \tilde{\Theta}_{k,n,p}^T$ is determined only by the pulse-shaping filter $h(n)$ and can be approximately considered nonzero only when $|l-k| \leq 1$ and $|m-n| \leq 1$ for any $p = 0, 1, \ldots, L_g - 1$. For example, with the IOTA filter, some values of $\Theta_{l,m}^* \tilde{\Theta}_{k,n,0}^T$ are shown in Table 8.6. Therefore, the data symbols $a_k(m)$, $m = 2, 3, \ldots, 2N_d - 2$, have almost no interference to pilot symbols $a_k(0)$. By making $a_k(1) = a_k(2N_d - 1) = 0$, $k = 0, 1, \ldots, K-1$, $\mathbf{s}^T \tilde{\Theta}_{k,0,p}^T$ is determined only by the pilots.

$$\mathbf{s}^T \tilde{\Theta}_{k,0,p}^T \approx \sum_{l=0}^{K-1} \Theta_{l,0}^* \tilde{\Theta}_{k,0,p}^T a_l(0). \qquad (8.68)$$

Therefore, according to Eqs. (8.66), (8.68), $\Theta \tilde{\mathbf{S}}$ can be approximately obtained, where its $(k+1)$th row $\Theta_{k,0} \tilde{\mathbf{S}}$ can be written as

$$\Theta_{k,0} \tilde{\mathbf{S}} \approx \left[\sum_{l=0}^{K-1} \Theta_{l,0}^* \tilde{\Theta}_{k,0,0}^T a_l(0), \sum_{l=0}^{K-1} \Theta_{l,0}^* \tilde{\Theta}_{k,0,1}^T a_l(0), \right.$$
$$\left. \cdots, \sum_{l=0}^{K-1} \Theta_{l,0}^* \tilde{\Theta}_{k,0,L-1}^T a_l(0) \right] (\mathbf{W}^H \mathbf{W})^{-1} \mathbf{W}^H. \qquad (8.69)$$

For simplicity, $\Theta \tilde{\mathbf{S}}$ is treated as a known matrix in the following. For the channel estimation model in Eq. (8.65), the noise variance matrix is

$$\tilde{\mathbf{V}} = \mathrm{E}\left[\Theta \boldsymbol{\eta}(\Theta \boldsymbol{\eta})^H\right] - \mathrm{E}\left[\Theta \boldsymbol{\eta}\right] \mathrm{E}\left[(\Theta \boldsymbol{\eta})^H\right] = \Theta \Theta^H \sigma^2. \qquad (8.70)$$

Due to the noise correlation in Eq. (8.70), WLS method can be employed to obtain the channel estimation by minimizing WLSE,

$$\mathrm{WLSE} = \mathrm{E}\left[(\mathbf{y} - \Theta \tilde{\mathbf{S}} \mathbf{G})^H \tilde{\mathbf{V}}^{-1} (\mathbf{y} - \Theta \tilde{\mathbf{S}} \mathbf{G})\right]. \qquad (8.71)$$

The channel estimation of $\mathbf{G}$ can then be written as

$$\hat{\mathbf{G}} = \left(\tilde{\mathbf{s}}^H \Theta^H \tilde{\mathbf{v}}^{-1} \Theta \tilde{\mathbf{s}}\right)^{-1} \tilde{\mathbf{s}}^H \Theta^H \tilde{\mathbf{v}}^{-1} \mathbf{y}. \qquad (8.72)$$

8.3.3 POP Channel Estimation in CP-OQAM/FBMC Systems

Different from the WLS estimator, the POP requires a pair of pilots (i.e., $a_k(m), m = 0, 1, k = 0, 1, \ldots, K - 1$), which are determined by the data symbols, and does not require the insertion of zero symbols. As shown in Fig. 8.14, a pilot design module is required in the POP estimator to make some selected entries of the signal before the IDFT module equal 1 and after the DFT module, the received signals at the corresponding positions are the estimated channel frequency responses. Therefore, for the POP, the receiver for the channel estimation is very simple.

As shown in Fig. 8.14, at the transmitter, the signal before the N_c-point IDFT module is

$$\boldsymbol{\zeta} \triangleq \sum_{k=0}^{K-1} \mathbf{P}_k \mathbf{a}_k = [\zeta(0), \zeta(1), \ldots, \zeta(N_c - 1)]^T. \tag{8.73}$$

Then, at the receiver, after the N_c-point DFT module, the received signal is

$$\mathbf{r}_f = \text{diag}(\mathbf{G}) \sum_{k=0}^{K-1} \mathbf{P}_k \mathbf{a}_k + \mathbf{W}_{N_c} \boldsymbol{\eta},$$

$$= \begin{bmatrix} G_0 & & & \\ & G_1 & & \\ & & \ddots & \\ & & & G_{N_c-1} \end{bmatrix} \begin{bmatrix} \zeta(0) \\ \zeta(1) \\ \vdots \\ \zeta(N_c - 1) \end{bmatrix} + \mathbf{W}_{N_c} \boldsymbol{\eta}, \tag{8.74}$$

which is due to the diagonalization of the channel for the CP-OQAM/FBMC as explained before similar to OFDM systems. Let $r_f(k)$ denote the $(k + 1)$th entry of $\mathbf{r}_f$. Obviously, when $\zeta(k) = 1$, $r_f(k)$ is

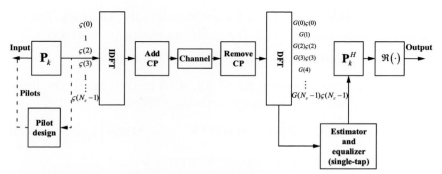

Fig. 8.14 POP channel estimator for CP-OQAM/FBMC systems.

the estimation of H_k. Thus, the key idea of POP is to design the pilot symbols $a_k(m), m = 0, 1$, to make $\zeta(k) = 1$. To be able to covert the channel frequency response to time channel impulse response, let us choose $k = 0, N_d, \ldots, (K-1)N_d$ [25], where K is the number of the selected $\zeta(k) = 1$, N_d is the interval size of the selected $\zeta(k) = 1$. Let $p_{l,k,m}$ denote the (k, m)th entry of $\mathbf{P}_l$ and $a_k(m), m = 0, 1, k = 0, 1, \ldots, K-1$, be the pilot symbols. Thus, $\zeta(k)$ is written as

$$\zeta(k) = \sum_{l=0}^{K-1} \sum_{m=0}^{1} p_{l,k,m} a_l(m) + \sum_{l=0}^{K-1} \sum_{m=2}^{2N_d-1} p_{l,k,m} a_l(m) = 1,$$
$$k = 0, N_d, \ldots, (K-1)N_d. \tag{8.75}$$

Since $\mathbf{P}_l$ is a complex matrix, we have

$$\Re \left\{ \sum_{l=0}^{K-1} \sum_{m=0}^{1} p_{l,k,m} a_l(m) + \sum_{l=0}^{K-1} \sum_{m=2}^{2N_d-1} p_{l,k,m} a_l(m) \right\} = 1,$$
$$\Im \left\{ \sum_{l=0}^{K-1} \sum_{m=0}^{1} p_{l,k,m} a_l(m) + \sum_{l=0}^{K-1} \sum_{m=2}^{2N_d-1} p_{l,k,m} a_l(m) \right\} = 0,$$
$$k = 0, N_d, \ldots, (K-1)N_d. \tag{8.76}$$

Let $\boldsymbol{\theta} = [\theta(0), \theta(1), \ldots, \theta(2K-1)]^T$ with

$$\begin{cases} \theta(2i) = 1 - \Re \left\{ \sum_{l=0}^{K-1} \sum_{m=2}^{2N_d-1} p_{l,iN_d,m} a_l(m) \right\} \\ \theta(2i+1) = -\Im \left\{ \sum_{l=0}^{K-1} \sum_{m=2}^{2N_d-1} p_{l,iN_d,m} a_l(m) \right\}, \end{cases}$$

$\mathbf{a}_p = [a_0(0), a_0(1), a_1(0), a_1(1), \ldots, a_{K-1}(0), a_{K-1}(1)]^T$, as the pilot symbol vector. Let $\boldsymbol{\Psi}$ be a $2K \times 2K$ matrix with the $(2i, 2l)$th entry $\Re\{p_{l,iN_d,0}\}$, the $(2i, 2l+1)$th entry $\Re\{p_{l,iN_d,1}\}$, the $(2i+1, 2l)$th entry $\Im\{p_{l,iN_d,0}\}$, and $(2i+1, 2l+1)$th entry $\Im\{p_{l,iN_d,1}\}$. Eq. (8.76) can be rewritten as

$$\boldsymbol{\theta} = \boldsymbol{\Psi} \mathbf{a}_p. \tag{8.77}$$

Based on the least square criterion, the transmitted pilot can be obtained

$$\mathbf{a}_p = \boldsymbol{\Psi}^\dagger \boldsymbol{\theta}, \tag{8.78}$$

where $\boldsymbol{\Psi}^\dagger$ is the generalized inverse matrix of $\boldsymbol{\Psi}$. With the designed pilots, the selected $\zeta(k), k = 0, N_d, \ldots, (K-1)N_d$, are 1 and the estimation of G_k at the receiver can be directly given

$$\hat{G}_k = r_f(k), \quad k = 0, N_d, \ldots, (K-1)N_d. \tag{8.79}$$

Since $G_k = \sum_{l=0}^{L_g-1} g(l)e^{-j2\pi kl/N_c}$, $k = 0, N_d, \ldots, (K-1)N_d$, the estimation of the time domain channel can be obtained

$$\hat{\mathbf{g}} = (\mathbf{B}^H\mathbf{B})^{-1}\mathbf{B}^H\tilde{\mathbf{r}}, \tag{8.80}$$

where $\mathbf{B}$ is a $K_p \times L$ matrix with the (m, n)th entry $[\mathbf{B}]_{m,l} = e^{-j2\pi mN_d l/N_c}$ and $\tilde{\mathbf{r}} = [r_f(0), r_f(N_d), \ldots, r_f((K-1)N_d)]^T$. Let $\hat{g}_m$ denote the $(m+1)$th entry of $\hat{\mathbf{g}}$. Finally, the frequency domain channel estimation is obtained by the N_c-point DFT transform

$$\hat{G}_k = \sum_{m=0}^{L_g-1} \hat{h}_m e^{-j2\pi mk/N_c}, \quad k = 0, 1, \ldots, N_c - 1, \tag{8.81}$$

which is employed for the single-tap equalizer as Eq. (8.60).

8.3.4 Simulation Results

In this section, numerical simulations are performed to evaluate the performances of the channel estimators in CP-OQAM/FBMC systems. For comparison, the performance of IAM [6] is also given. In the simulations, we consider a quasistatic channel model with $L_g = 30$ sample-spaced independent Rayleigh fading paths, which has an exponential power delay profile as: $\mu(l) = e^{-\tilde{\beta}l}$, where $\tilde{\beta} = 0.2$ and $l = 0, 1, \ldots, L_g - 1$. The system model and parameters are listed as follows:

- Subcarrier number: K=128
- Modulation: 4QAM
- One block: 128×20 real symbols
- Length of CP: 32 samples
- Pulse-shaping filter: IOTA
- Channel coding: quasicyclic LDPC codes (QC-LDPC) with code length 1152 and code rate $\frac{1}{2}$.

Figs. 8.15 and 8.16 show the BER performances of the WLS and POP estimators in the CP-OQAM/FBMC system. Both of the WLS and POP estimators achieve good BER performances. It is worth mentioning that compared with the WLS estimator, the receiver of the POP is simple. However, the design of pilots of the POP estimator requires more computation at the transmitter. In addition, CP-OQAM/FBMC systems exhibit a significant performance gain over OQAM/FBMC systems with single-tap equalizer, which indicates that the conventional OQAM/FBMC systems cannot efficiently fight against the multipath channel without the

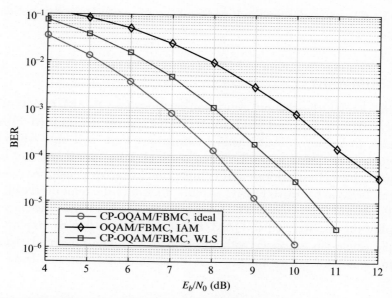

Fig. 8.15 BER performance of WLS in OQAM/FBMC systems.

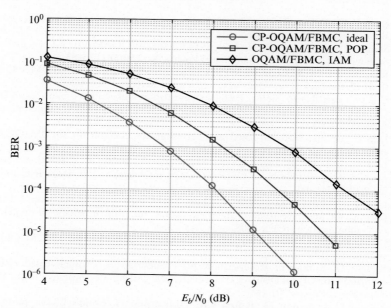

Fig. 8.16 BER performances of the POP in CP-OQAM/FBMC systems.

help of CP, especially for the broadband systems with large channel delay spread.

8.4 SUMMARY

In this chapter, the channel estimation methods for OQAM/FBMC were presented, as well as in CP-OQAM/FBMC systems. In OQAM/FBMC systems, frequency domain channel estimations work well only when the channel delay spread is small. When channel delay spread is large, time domain channel estimations and complex multiple-tap equalizers have to be employed in OQAM/FBMC systems. To reduce the pilot overhead, the CAP method is presented, which can achieve good spectral efficiency. To simplify the receiver, CP-OQAM/FBMC systems enable to employ the simple single-tap equalizer with the help of CP. The presented channel estimations can work well even when the channel delay spread is large.

REFERENCES

[1] Kong D, Qu D, Jiang T. Time domain channel estimation for OQAM-OFDM systems: algorithms and performance bounds. IEEE Trans Signal Process 2014;63(2):322–30.
[2] Siohan P, Siclet C, Lacaille N. Analysis and design of OQAM-OFDM systems based on filterbank theory. IEEE Trans Signal Process 2002;50(5):1170–83.
[3] Javaudin JP, Lacroix D, Rouxel A. Pilot-aided channel estimation for OFDM/OQAM. In: IEEE semiannual on vehicular technology conference (VTC); 2003.
[4] Lélé C, Legouable R, Siohan P. Channel estimation with scattered pilots in OFDM/OQAM. In: IEEE workshop on signal processing advances in wireless communications (SPAWC); 2008.
[5] Lélé C. Iterative scattered-based channel estimation method for OFDM/OQAM. EURASIP J Adv Signal Process 2012;42.
[6] Lélé C, Javaudin JP, Legouable R, Skrzypczak A, Siohan P. Channel estimation methods for preamble-based OFDM/OQAM modulations. Eur Trans Telecommun 2007;19(7):741–50.
[7] Lélé C, Siohan P, Legouable R. 2 dB better than CP-OFDM with OFDM/OQAM for preamble-based channel estimation. In: IEEE international conference on communication (ICC); 2008.
[8] Du J, Signell S. Novel preamble-based channel estimation for OFDM/OQAM systems. In: IEEE international conference on communication; 2009.
[9] Katselis D, Kofidis E, Rontogiannis A, Theodoridis S. Preamble-based channel estimation for CP-OFDM and OFDM/OQAM systems: a comparative study. IEEE Trans Signal Process 2010;58(5):2911–6.
[10] Katselis D, Bengtsson M, Rojasa CR, Hjalmarsson H, Kofidis E. On preamble-based channel estimation in OFDM/OQAM systems. In: European signal processing conference; 2011.
[11] Lin H, Siohan P. Robust channel estimation for OFDM/OQAM. IEEE Commun Lett 2009;13(10):724–6.

[12] Hu S, Wu G, Li S. Preamble design and iterative channel estimation for OFDM/offset QAM system. J Netw 2009;4(10):963–72.

[13] Katselis D, Rojas CR, Bengtsson M, Hjalmarsson H. Frequency smoothing gains in preamble-based channel estimation for multicarrier systems. Signal Process 2013;93(9):2777–82.

[14] Kofidis E, Katselis D, Rontogiannis A, Theodoridi S. Preamble-based channel estimation in OFDM/OQAM systems: a review. Signal Process 2013;93(7):2038–54.

[15] Gao X, Wang W, Xia XG, Au EKS, You X. Cyclic prefixed OQAM-OFDM and its application to single-carrier FDMA. IEEE Trans Commun 2011;59(5):1467–80.

[16] Du J, Signell S. Time frequency localization of pulse shaping filter in OFDM/OQAM systems. In: International conference on information communication and signal processing (ICICS); 2007.

[17] Siohan P, Roche C. Cosine-modulated filterbanks based on extended Gaussian function. IEEE Trans Signal Process 2000;48(11):3052–61.

[18] Farhang-Boroujeny B. OFDM versus filter bank multicarrier. IEEE Signal Process Mag 2011;28(3):92–112.

[19] Mirabbasi S, Martin K. Overlapped complex-modulated transmultiplexer filters with simplified design and superior stopbands. IEEE Trans Circuits Syst II Analog Digit Signal Process 2003;50(8):456–69.

[20] Bellanger MG. Specification and design of a prototype filter for filter bank based multicarrier transmission. In: IEEE international conference on acoustics, speech, and signal processing; 2001.

[21] Viholainen A, Bellanger M, Huchard M. Prototype filter and structure optimization, 2009. Available from: http://www.ict-phydyas.org/delivrables/PHYDYAS-D5-1.pdf/view.

[22] Bertsekas DP. Nonlinear programming. Cambridge, MA: Athena Scientific; 1999.

[23] The IEEE 802.16 Broadband Wireless Access Working Group, Channel Models for Fixed Wireless Applications, 2001. Available from: http://www.ieee802.org/16/tg3/contrib/802163c-01_29r4.pdf.

[24] Fazel K, Kaiser S. Multi-carrier and spread spectrum systems: from OFDM and MC-CDMA to LTE and WiMAX. New York, NY: John Wiley & Sons; 2008.

[25] Kong D, Xia XG, Jiang T, Gao X. Channel estimation in CP-OQAM-OFDM systems. IEEE Trans Signal Process 2014;62(21):5775–86.

CHAPTER 9

MIMO OQAM/FBMC

Multiple-input and multiple-output (MIMO) techniques can greatly improve the spectral efficiency and energy efficiency of offset quadrature amplitude modulation-based filter bank multicarrier (OQAM/FBMC) systems. Therefore, MIMO OQAM/FBMC is a very promising potential physical layer technique for future wireless communication systems. However, compared the OFDM, the combination of OQAM/FBMC to MIMO is more complicated due to the intrinsic imaginary interference in OQAM/FBMC systems [1]. The two main approaches of the MIMO concept, namely spatial multiplexing and spatial diversity [2], can be applied to the OQAM/FBMC as described following.

Spatial Multiplexing. In spatial multiplexing, the system throughput can be increased by transmitting different data streams over different antennas. The case of two transmit antennas and two receive antennas is illustrated in Fig. 9.1 [3].

The channel matrix consists of four elementary channels and can be modeled as

$$G = \begin{bmatrix} g_{11} & g_{21} \\ g_{12} & g_{22} \end{bmatrix}. \tag{9.1}$$

In the following, the signal detection of the MIMO OQAM/FBMC systems is described in Fig. 9.2. For a particular symbol and a particular subcarrier, the signal input at the first receive antenna is written as

$$x_1 = (d_1 + ju_1)g_{11} + (d_2 + ju_2)g_{21}, \tag{9.2}$$

where d_1 and d_2 are the data symbols at the first and the second transmit antennas, respectively, and u_1, u_2 are the corresponding imaginary interference. The complex scalars g_{11} and g_{21} represent the channel between the transmit antennas and the first receive antenna, assuming perfect frequency and time synchronization.

OQAM/FBMC for Future Wireless Communications
http://dx.doi.org/10.1016/B978-0-12-813557-0.00009-7

© 2018 Elsevier Ltd.
All rights reserved.   213

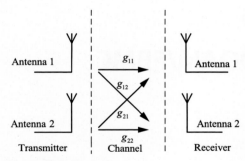

Fig. 9.1 MIMO 2 × 2 transmissions.

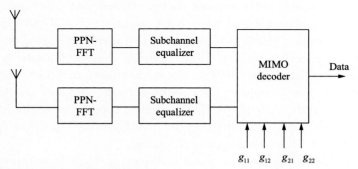

Fig. 9.2 MIMO 2 × 2 receiver structure.

Similarly, the signal at the second receive antenna is obtained as

$$x_2 = (d_1 + ju_1)g_{12} + (d_2 + ju_2)g_{22}. \tag{9.3}$$

From Eqs. (9.2), (9.3), the transmit data symbols can be recovered theoretically by the following zero-forcing (ZF) algorithm

$$\begin{bmatrix} d_1 \\ d_2 \end{bmatrix} = \mathrm{Re}\left[\begin{bmatrix} g_{11} & g_{21} \\ g_{12} & g_{22} \end{bmatrix}^{-1} \begin{bmatrix} x_1 \\ x_2 \end{bmatrix} \right]. \tag{9.4}$$

In the presence of additive white Gaussian noise (AWGN) components n_1 and n_2 with variance σ^2, the minimum mean square error (MMSE) algorithm yields the following estimation:

$$\begin{bmatrix} \tilde{d}_1 + j\tilde{u}_1 \\ \tilde{d}_2 + j\tilde{u}_2 \end{bmatrix} = FG \begin{bmatrix} d_1 + ju_1 \\ d_2 + ju_2 \end{bmatrix} + F \begin{bmatrix} n_1 \\ n_2 \end{bmatrix}, \tag{9.5}$$

with

$$F = \left(G^H G + \sigma^2 I \right)^{-1} G^H. \tag{9.6}$$

Then, by the operation of taking real part, the data symbols can be recovered. Note that the MMSE algorithm can achieve better system performance than ZF at the cost of high complexity.

Spatial Diversity. Diversity can be implemented in the time or frequency domain at the transmitter or the receiver. In the following, we will take transmit diversity as example to introduce spatial diversity. Due to the presence of the imaginary interference term, the OQAM/FBMC introduces specific constraints. Thus, to achieve full diversity, the interference term has to be separated from the data in the diversity processing. The case of two transmit antennas and one receive antenna is considered and illustrated in Fig. 9.3 [4]. In this case, the MIMO 2 × 2 channel is modeled by a two-tap FIR filter, as shown in Fig. 9.4.

The sequence of data $d(m)$ is assumed to be real and, taking into account the interference sequence $u(m)$, the received signal is written as

$$x(m) = g_1(d(m) + ju(m)) + g_2(d(m-1) + ju(m-1)). \qquad (9.7)$$

Since the filter coefficients g_1 and g_2 are assumed to be complex scalars, the real data and the imaginary interference samples can be separated by the conjugate of the channel filter. The received signal is then given by

$$y(m+1) = g_1^* x(m+1) + g_2^* x(m), \qquad (9.8)$$

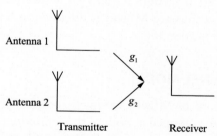

Fig. 9.3 Transmit diversity with two antennas.

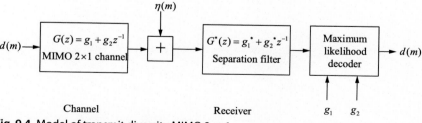

Fig. 9.4 Model of transmit diversity MIMO 2 × 1.

or

$$y(m+1) = |g_1|^2(d(m+1) + ju(m+1))$$
$$+ [g_1g_2^\star + g_2g_1^\star](d(m) + ju(m)) + |g_2|^2(d(m-1) + ju(m-1)).$$
$$(9.9)$$

Then, a sequence free of interference is obtained by

$$\text{Re}(y(m+1)) = |g_1|^2 d(m+1) + 2\text{Re}(g_1g_2^\star)d(m) + |g_2|^2 d(m-1). \quad (9.10)$$

The data are then retrieved through decision feedback equalization or maximum likelihood (ML) decoding. However, it is worthwhile to point out that the noise component in OQAM/FBMC is correlated and the ML technique yields suboptimal results. Therefore, the noise has to be included in the calculation of the equivalent signal-to-noise ratio (SNR) of the ML decoder and iterative techniques must be employed to reach a value close to the full diversity SNR which is expressed by

$$\text{SNR}_{\text{div}} = \frac{|g_1|^2 + |g_2|^2}{\sigma^2}. \quad (9.11)$$

The rest of this chapter is organized as follows. Section 9.1 describes the MIMO channel model. The MIMO OQAM/FBMC system is briefly discussed in Section 9.2. Then, a transmit diversity scheme and a beamforming scheme are introduced in Sections 9.3 and 9.4, respectively. The application of OQAM/FBMC to massive MIMO is discussed in Section 9.5. Finally, summary is made in Section 9.6.

9.1 MIMO CHANNEL MODEL

The principle diagram of MIMO system is shown in Fig. 9.5 [5] where multiple antennas are equipped at both of the transmitter and receiver. The signal at the transmit antenna array can be expressed as

$$\mathbf{x}(t) = [x_1(t) \ x_2(t) \ \cdots \ x_{N_T}(t) \ ]^T, \quad (9.12)$$

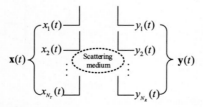

Fig. 9.5 The diagram of MIMO system.

where $[\cdot]^T$ denotes transpose of a vector or matrix and $x_i(t)$ is the signal at the ith transmit antenna.

The signal at the receive antenna array can be expressed as

$$\mathbf{y}(t) = [\, y_1(t) \;\; y_2(t) \;\; \cdots \;\; y_{N_T}(t) \,]^T, \tag{9.13}$$

where $y_i(t)$ is the signal at the ith receive antenna.

9.1.1 Non-Frequency-Selective Channel Model

For the non-frequency-selective fading channel, MIMO channel model is relatively simple due to the fact that the interantenna subcarrier can be equivalent to a Rayleigh fading channel. Each subcarrier of the MIMO channel model can be modeled as

$$g_{j,i}(t, \tau) = g_{j,i}(t)\delta(\tau - \tau_0), \tag{9.14}$$

where $i = 1, \ldots, N_T; j = 1, \ldots, N_R$. $\left| g_{j,i}(t) \right|$ obeys the Rayleigh distribution.

Assume that the MIMO channel matrix can be expressed as

$$\mathbf{G} = \begin{pmatrix} g_{11} & g_{12} & \cdots & g_{1N_T} \\ g_{21} & g_{22} & \cdots & g_{2N_T} \\ \vdots & \vdots & \ddots & \vdots \\ g_{N_R 1} & g_{N_R 2} & \cdots & g_{N_R N_T} \end{pmatrix}. \tag{9.15}$$

The received signal in matrix form can then be written as

$$\mathbf{Y} = \mathbf{G}\,\mathbf{X} + \boldsymbol{\eta}, \tag{9.16}$$

where $\boldsymbol{\eta}$ is Gaussian white noise matrix with zero mean.

9.1.2 Frequency Selective Channel Model

For the frequency selective fading channels, the MIMO channel matrix is expressed as

$$\mathbf{G}\,(\tau) = \sum_{l=1}^{L} \mathbf{G}^{\,i}\delta(\tau - \tau_l), \tag{9.17}$$

where $\mathbf{G}\,(\tau) \in C_{N_R \times N_T}$ and

$$\mathbf{G}^{\,l} = \begin{pmatrix} g_{11}^{l} & \cdots & g_{1N_T}^{l} \\ \vdots & \ddots & \vdots \\ g_{N_R 1}^{l} & \cdots & g_{N_R N_T}^{l} \end{pmatrix}_{N_R \times N_T}. \tag{9.18}$$

$\mathbf{G}^l$ is a complex-valued matrix which denotes the linear transformation between two antenna arrays when the time delay is τ; $g_{j,i}^l$ is the complex-valued transmission coefficient from ith transmit antenna to jth receive antenna.

The relationship between channel output vector $\mathbf{y}(t)$ and input vector $\mathbf{x}(t)$ can be expressed as

$$\mathbf{y}(t) = \int \mathbf{G}(\tau)\mathbf{x}(t - \tau)d\tau. \tag{9.19}$$

The MIMO channel model in Eq. (9.20) can be regard as the extended version of single-input single-output channel standard model. However, the main difference is that the tap coefficient of channel model is not just a simple scalar, but a matrix. The size of the matrix is related to the number of antennas at both ends of MIMO system.

By sampling Eq. (9.20), the discrete model of the frequency selective channel can be expressed as

$$\mathbf{y}_n = \sum_{l=0}^{L} \mathbf{G}^l \mathbf{x}_{n-l} + \boldsymbol{\eta}_n, \tag{9.20}$$

where $\boldsymbol{\eta}_n$ is the Gaussian white noise matrix with zero mean. In MIMO channel analysis and estimation, we can use this formula to simplify calculation.

9.2 MIMO OQAM/FBMC SYSTEMS

Communication systems employing multiple antennas at both the transmitter and receiver, which can increase the system throughput and the reliability of the link, have become very popular in the past decades and have been included in most of the recent standards. The performance enhancements mainly come from the array gain, diversity gain and spatial multiplexing gain [3].

- *Array gain.* Array gain can be obtained through processing at the transmitter and the receiver, which results in an increase in average receive SNR due to a coherent combining effect. The effect of the array gain depends on the number of transmit and receive antennas. In addition, transmit/receive array gain requires channel knowledge in the transmitter and receiver, respectively.

- *Diversity gain.* The wireless channel fluctuates randomly, which usually causes the fading of the transmitted signal. Diversity is a powerful technique to mitigate fading in wireless communication systems. Diversity gain is obtained by transmitting the signal over multiple independently fading paths (in time/frequency/space). It is possible to exploit spatial diversity gain in the absence of channel knowledge at the transmitter by using suitably designed transmit signals. The corresponding technique is known as space–time coding (STC).
- *Spatial multiplexing gain.* MIMO channels offer a linear increase in capacity without additional power or bandwidth. This gain is referred as spatial multiplexing gain. The spatial multiplexing gain is realized by transmitting independent data signals from individual antennas. Under conducive channel conditions, such as rich scattering, the receiver can separate the different streams, yielding a linear increase in capacity.

To improve the performance, it is natural to consider combining OQAM/FBMC with MIMO. However, the combination is very complicated due to the intrinsic imaginary interference in OQAM/FBMC [6]. The equivalent SISO baseband OQAM/FBMC signal can be expressed as

$$s(k) = \sum_{m=0}^{K-1} \sum_{n \in \mathbb{Z}} a_m(n) \underbrace{h\left[k - n\frac{K}{2}\right] e^{j2\pi mk/K} e^{j(m+n)\pi/2}}_{h_{m,n}[k]}, \qquad (9.21)$$

where K is the number of subcarriers and k is the sample index. The transmitted data symbol $a_m(n)$ is real-valued with frequency index m and time index n. $a_m(2k)$ and $a_m(2k+1)$ are obtained by taking the real and imaginary parts of a complex-valued symbol from QAM constellation, respectively. T is the complex symbol interval, and $T/2$ is the real-valued symbol interval. $h(n)$ is a symmetrical real-valued prototype filter.

Let us consider a locally time-invariant channel, and a neighborhood $\Omega_{\Delta m, \Delta n}$ of around the (m, n) position is defined as

$$\Omega_{\Delta m, \Delta n} = \left\{ (p, q), |p| \leq \Delta m, |q| \leq \Delta n \,|\, G_{m+p, n+q} \approx G_{m,n} \right\}, \qquad (9.22)$$

where $G_{m,n}$ is the channel frequency response on subcarrier m at time index n. We could also define $\Omega_{\Delta m, \Delta n}^{\star} = \Omega_{\Delta m, \Delta n} - \{(0, 0)\}$ [6].

We assume that the symbol interval is much longer than Δ. Therefore, $h[k] \approx h[k+\tau]$, for all $\tau \in [0, \Delta]$. The demodulated signal after the analysis filter bank could then be expressed as

$$\gamma_m(n) \approx G_{m,n} \left(a_m(n) + ja_{m,n}^{(i)} \right), \tag{9.23}$$

where $ja_{m,n}^{(i)}$, the intrinsic imaginary interference, is given as

$$ja_{m,n}^{(i)} = \sum_{(p,q)\in\Omega_{\Delta m,\Delta n}^\star} a_{m+p}(n+q) \langle h_{m+p,n+q}, h_{m,n} \rangle$$

$$= \sum_{(p,q)\in\Omega_{\Delta m,\Delta n}^\star} a_{m+p}(n+q)$$

$$\underbrace{\sum_{k=-\infty}^{+\infty} h\left[k - (n+q)\frac{K}{2}\right] h\left[k - n\frac{K}{2}\right] e^{j2\pi pk/K} e^{j(p+q)\pi/2}}_{c_{p,q}^n}, \tag{9.24}$$

where $\langle \cdot, \cdot \rangle$ denotes the inner product.

The fundamental characteristics of $c_{p,q}^n$ can be obtained as

$$c_{p,q}^n = (-1)^{pn} c_{p,q}^0,$$

$$c_{-p,q}^0 = (-1)^{p(q+1)} c_{p,q}^0,$$

$$c_{-p,q}^n = (-1)^{q+1} c_{p,q}^n. \tag{9.25}$$

Due to the intrinsic imaginary interference in OQAM/FBMC, OQAM/FBMC cannot be directly combined with MIMO. So far, many works focus on exploiting spatial multiplexing and diversity in MIMO OQAM/FBMC [6–10].

9.3 STBC AND SFBC

The STC concept was first presented in [11]. Then, a key development of the STC concept was revealed in [12], which is called space-time trellis coding. The symbols are encoded according to the antennas through which they are simultaneously transmitted, and are decoded using an ML decoder. This scheme can provide considerable performance gains by combining the benefits of forward error correction coding and diversity transmission. However, this scheme requires a multidimensional Viterbi algorithm at the receiver for decoding. Considering the large decoding complexity, it is not practical for application. Hence, Alamouti proposed a space-time block coding (STBC) scheme with small computation complexity in [4].

9.3.1 STBC and SFBC in OFDM

In general, a code $\mathcal{C}$ is a 2×2 matrix with columns corresponding to antennas and rows corresponding to either time indexes for STBC or frequency tones for space-frequency block code (SFBC). The elements of $\mathcal{C}$ are linear combinations of S_1, S_2 and their conjugates. The code $\mathcal{C}$ provides a full diversity gain of two if it satisfies the orthogonality condition of the Alamouti code [13, 14] as

$$\mathcal{C}^H \mathcal{C} = \left(|S_1|^2 + |S_2|^2 \right) \mathbf{I}, \tag{9.26}$$

where $\mathbf{I}$ is a 2×2 identity matrix. For example, in a conventional SFBC-OFDM system, there exists

$$\mathcal{C} = \left[\begin{array}{cc} S_1 & S_2 \\ -S_2^\star & S_1^\star \end{array} \right]. \tag{9.27}$$

Obviously, Eq. (9.27) satisfies Eq. (9.26).

Supposing the channel frequency response does not vary over the subcarriers of the code, the received signals could be expressed in the frequency domain as

$$\left[\begin{array}{c} R_1 \\ R_2^\star \end{array} \right] = \underbrace{\left[\begin{array}{cc} G_1 & G_2 \\ G_2^\star & -G_1^\star \end{array} \right]}_{\bar{\mathbf{G}}} \left[\begin{array}{c} S_1 \\ S_2 \end{array} \right] + \left[\begin{array}{c} \eta_1 \\ \eta_2^\star \end{array} \right], \tag{9.28}$$

where R_1 and R_2 are the received signals on the two subcarriers, G_1 and G_2 are the channel frequency responses of the two antennas, η is the AWGN. It is observed that the equivalent channel matrix $\bar{\mathbf{G}}$ is orthogonal with the help of the orthogonal code $\mathcal{C}$. The received signals could then be processed by a space-frequency combiner, which computes the decision variables as

$$
\begin{aligned}
\left[\begin{array}{c} \tilde{R}_1 \\ \tilde{R}_2 \end{array} \right] &= \bar{\mathbf{G}}^H \left[\begin{array}{c} R_1 \\ R_2^\star \end{array} \right] \\
&= \left[\begin{array}{cc} G_1^\star & G_2 \\ G_2^\star & -G_1 \end{array} \right] \left[\begin{array}{cc} G_1 & G_2 \\ G_2^\star & -G_1^\star \end{array} \right] \left[\begin{array}{c} S_1 \\ S_2 \end{array} \right] + \bar{\mathbf{G}}^H \left[\begin{array}{c} \eta_1 \\ \eta_2^\star \end{array} \right] \\
&= \left(|G_1|^2 + |G_2|^2 \right) \left[\begin{array}{c} S_1 \\ S_2 \end{array} \right] + \bar{\mathbf{G}}^H \left[\begin{array}{c} \eta_1 \\ \eta_2^\star \end{array} \right].
\end{aligned} \tag{9.29}
$$

The ML detection is then the optimum way to make the symbol decisions, and a full diversity gain of two is achieved.

9.3.2 A Block-Wise SFBC Scheme for OQAM/FBMC

The block-wise SFBC scheme for OQAM/FBMC systems is depicted in Fig. 9.6, which has a block size K. For the time index n, a block of real-valued data symbols is divided into an upper block and a lower block. Each of the two blocks has K subcarriers and they are isolated by a guard band. With the PHYDYAS prototype filter [15–17], where the interference mainly comes from the two adjacent subcarriers [18], the upper blocks could be well isolated from the lower blocks with one guard subcarrier.

As depicted in Fig. 9.6, the (k, n)th real SFBC code is designed as

$$C^{r}_{k,n} = \left[\begin{array}{cc} a_k(n) & b_k(n) \\ (-1)^{n+1}b_k(n) & (-1)^{n}a_k(n) \end{array} \right]. \tag{9.30}$$

Supposing the channel frequency response does not vary over the subcarriers of the code, the equivalent (k, n)th complex code is given as

$$C_{k,n} = \left[\begin{array}{cc} a_k(n) + jd^{(i)}_{m(k,1),n,1} & b_k(n) + jd^{(i)}_{m(k,1),n,2} \\ (-1)^{n+1}b_k(n) + jd^{(i)}_{m(k,2),n,1} & (-1)^{n}a_k(n) + jd^{(i)}_{m(k,2),n,2} \end{array} \right], \tag{9.31}$$

where the function $m(k, s)$ gives the subcarrier index of the (k, n)th SFBC code, with $s = 1, 2$ representing the upper block and the lower block, respectively (n is omitted in the function since it is not related to the subcarrier index), and $jd^{(i)}_{m(k,s),n,l}$ is the imaginary interference suffered by the data on the subcarrier $m(k, s)$ at the time n from the antenna l. Since the PHYDYAS prototype filter is employed in this chapter, we only need to consider the interference from the neighborhood $\Omega^{\star}_{1,r}$ (particularly, $r = 4$ is considered in our simulations due to that the interference outside $\Omega_{1,4}$ is very small).

We then have

$$jd^{(i)}_{m(k,1),n,1} = \sum_{(p,q)\in\Omega^{\star}_{1,r}} a_{k+p}(n + q)c^{n}_{-p,q}$$

$$= \sum_{(p,q)\in\Omega^{\star}_{1,r}} (-1)^{q+1} a_{k+p}(n + q)c^{n}_{p,q}, \tag{9.32}$$

$$jd^{(i)}_{m(k,1),n,2} = \sum_{(p,q)\in\Omega^{\star}_{1,r}} b_{k+p}(n + q)c^{n}_{-p,q}$$

$$= \sum_{(p,q)\in\Omega^{\star}_{1,r}} (-1)^{q+1} b_{k+p}(n + q)c^{n}_{p,q}. \tag{9.33}$$

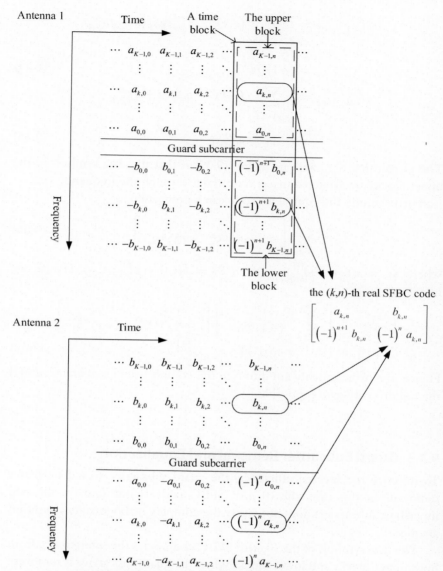

Fig. 9.6 The block-wise SFBC scheme for OQAM/FBMC systems.

$$jd^{(i)}_{m(k,2),n,1} = \sum_{(p,q)\in\Omega^\star_{1,r}} (-1)^{(n+q)+1} b_{k+p}(n+q) c^n_{p,q}$$

$$= (-1)^n jd^{(i)}_{m(k,1),n,2}, \tag{9.34}$$

$$jd^{(i)}_{m(k,2),n,2} = \sum_{(p,q)\in\Omega^\star_{1,r}} (-1)^{n+q} a_{k+p}(n+q) c^n_{p,q}$$

$$= (-1)^{n+1} jd^{(i)}_{m(k,1),n,1}. \tag{9.35}$$

The reason that $c^n_{-p,q}$ appears in Eq. (9.32) is that the data symbols in the upper blocks are arranged in a reverse order through the frequency axis.

Furthermore, the equivalent code can be rewritten as

$$C_{k,n} = \begin{bmatrix} S_1 & S_2 \\ (-1)^{n+1} S_2^\star & (-1)^n S_1^\star \end{bmatrix}, \tag{9.36}$$

where $S_1 = a_k(n) + jd^{(i)}_{m(k,1),n,1}$ and $S_2 = b_k(n) + jd^{(i)}_{m(k,1),n,2}$. Hence, we have

$$C^H_{k,n} C_{k,n} = \begin{bmatrix} S_1^\star & (-1)^{n+1} S_2 \\ S_2^\star & (-1)^n S_1 \end{bmatrix} \begin{bmatrix} S_1 & S_2 \\ (-1)^{n+1} S_2^\star & (-1)^n S_1^\star \end{bmatrix}$$

$$= \left(|S_1|^2 + |S_2|^2\right) \mathbf{I}. \tag{9.37}$$

Hence, the orthogonality condition in the complex field is guaranteed and the full diversity gain is possible.

9.3.3 Guard Subcarrier Removal and Equalization

To improve the spectral efficiency, the guard subcarrier would be better removed. In this case, the $(0, n)$th SFBC code transmitted on the two innermost subcarriers does not satisfy the complex orthogonality condition anymore.

The data symbols of the $(0, n)$th SFBC code suffer the interference from both upper blocks and lower blocks when the guard subcarrier is removed. Thus, we divide the imaginary interference into two parts as

$$jd^{(i)}_{m(0,s),n,l} = \overline{jd}^{(i)}_{m(0,s),n,l} + j\underline{d}^{(i)}_{m(0,s),n,l}, \tag{9.38}$$

where $\overline{jd}^{(i)}_{m(0,s),n,l}$ represents the interference from upper blocks and $j\underline{d}^{(i)}_{m(0,s),n,l}$ represents the interference from lower blocks. According to Eq. (9.31), the equivalent $(0, n)$th complex code is rewritten as

$$\mathcal{C}_{0,n} =$$

$$\begin{bmatrix} a_0(n) + j\bar{d}^{(i)}_{m(0,1),n,1} + j\underline{d}^{(i)}_{m(0,1),n,1} & b_0(n) + j\bar{d}^{(i)}_{m(0,1),n,2} + j\underline{d}^{(i)}_{m(0,1),n,2} \\ (-1)^{n+1}b_0(n) + j\bar{d}^{(i)}_{m(0,2),n,1} + j\underline{d}^{(i)}_{m(0,2),n,1} & (-1)^{n}a_0(n) + j\bar{d}^{(i)}_{m(0,2),n,2} + j\underline{d}^{(i)}_{m(0,2),n,2} \end{bmatrix}. \tag{9.39}$$

For data symbols of the upper blocks, the interference is given as

$$j\bar{d}^{(i)}_{m(0,1),n,1} = \sum_{\substack{p\in[0,1],q\in[-r,r] \\ (p,q)\neq(0,0)}} a_p(n+q)c^n_{-p,q}$$

$$= \sum_{\substack{p\in[0,1],q\in[-r,r] \\ (p,q)\neq(0,0)}} (-1)^{q+1}a_p(n+q)c^n_{p,q}, \tag{9.40}$$

$$j\underline{d}^{(i)}_{m(0,1),n,1} = \sum_{\substack{p=-1 \\ q\in[-r,r]}} (-1)^{n+q+1}b_0(n+q)c^n_{-p,q}$$

$$= \sum_{\substack{p=1 \\ q\in[-r,r]}} (-1)^{n+q+1}b_0(n+q)c^n_{p,q}, \tag{9.41}$$

$$j\bar{d}^{(i)}_{m(0,1),n,2} = \sum_{\substack{p\in[0,1],q\in[-r,r] \\ (p,q)\neq(0,0)}} b_p(n+q)c^n_{-p,q}$$

$$= \sum_{\substack{p\in[0,1],q\in[-r,r] \\ (p,q)\neq(0,0)}} (-1)^{q+1}b_p(n+q)c^n_{p,q}, \tag{9.42}$$

$$j\underline{d}^{(i)}_{m(0,1),n,2} = \sum_{\substack{p=-1 \\ q\in[-r,r]}} (-1)^{n+q}a_0(n+q)c^n_{-p,q}$$

$$= \sum_{\substack{p=1 \\ q\in[-r,r]}} (-1)^{n+q}a_0(n+q)c^n_{p,q}. \tag{9.43}$$

For data symbols of the lower blocks, the interference is accordingly given as

$$j\bar{d}^{(i)}_{m(0,2),n,1} = \sum_{\substack{p=-1 \\ q\in[-r,r]}} a_0(n+q)c^n_{p,q}$$

$$= \sum_{\substack{p=1 \\ q\in[-r,r]}} a_0(n+q)c^n_{-p,q}$$

$$= \sum_{\substack{p=1 \\ q\in[-r,r]}} a_0(n+q)(-1)^{q+1}c_{p,q}^n$$

$$= (-1)^{n+1} \sum_{\substack{p=1 \\ q\in[-r,r]}} (-1)^{n+q}a_0(n+q)c_{p,q}^n$$

$$= (-1)^{n+1} j\underline{d}_{m(0,1),n,2}^{(i)}, \tag{9.44}$$

$$j\underline{d}_{m(0,2),n,1}^{(i)} = \sum_{\substack{p\in[0,1],q\in[-r,r] \\ (p,q)\neq(0,0)}} (-1)^{n+q+1}b_p(n+q)c_{p,q}^n$$

$$= (-1)^{n} \sum_{\substack{p\in[0,1],q\in[-r,r] \\ (p,q)\neq(0,0)}} (-1)^{q+1}b_p(n+q)c_{p,q}^n$$

$$= (-1)^{n} j\overline{d}_{m(0,1),n,2}^{(i)}, \tag{9.45}$$

$$j\overline{d}_{m(0,2),n,2}^{(i)} = \sum_{\substack{p=-1 \\ q\in[-r,r]}} b_0(n+q)c_{p,q}^n$$

$$= \sum_{\substack{p=1 \\ q\in[-r,r]}} b_0(n+q)c_{-p,q}^n$$

$$= \sum_{\substack{p=1 \\ q\in[-r,r]}} b_0(n+q)(-1)^{q+1}c_{p,q}^n$$

$$= (-1)^{n} \sum_{\substack{p=1 \\ q\in[-r,r]}} (-1)^{n+q+1}b_0(n+q)c_{p,q}^n$$

$$= (-1)^{n} j\underline{d}_{m(0,1),n,1}^{(i)}, \tag{9.46}$$

$$j\underline{d}_{m(0,2),n,2}^{(i)} = \sum_{\substack{p\in[0,1],q\in[-r,r] \\ (p,q)\neq(0,0)}} (-1)^{n+q}a_p(n+q)c_{p,q}^n$$

$$= (-1)^{n+1} \sum_{\substack{p\in[0,1],q\in[-r,r] \\ (p,q)\neq(0,0)}} (-1)^{q+1}a_p(n+q)c_{p,q}^n$$

$$= (-1)^{n+1} j\overline{d}_{m(0,1),n,1}^{(i)}. \tag{9.47}$$

According to Eqs. (9.40)–(9.47), Eq. (9.39) can be rewritten as

$$
\mathcal{C}_{0,n} =
$$
$$
\begin{bmatrix}
a_0(n) + j\overline{d}^{(i)}_{m(0,1),n,1} + j\underline{d}^{(i)}_{m(0,1),n,1} & b_0(n) + j\overline{d}^{(i)}_{m(0,1),n,2} + j\underline{d}^{(i)}_{m(0,1),n,2} \\
(-1)^{n+1}\left(b_0(n) - j\overline{d}^{(i)}_{m(0,1),n,2} + j\underline{d}^{(i)}_{m(0,1),n,2}\right) & (-1)^n\left(a_0(n) - j\overline{d}^{(i)}_{m(0,1),n,1} + j\underline{d}^{(i)}_{m(0,1),n,1}\right)
\end{bmatrix}.
$$
$$(9.48)$$

Supposing the channel frequency response does not vary over adjacent subcarriers, the received signals of the two innermost subcarriers after the analysis filter bank could be expressed in the frequency domain as

$$
\begin{aligned}
R^1_{0,n} &= G_1\left(a_0(n) + j\overline{d}^{(i)}_{m(0,1),n,1} + j\underline{d}^{(i)}_{m(0,1),n,1}\right) \\
&\quad + G_2\left(b_0(n) + j\overline{d}^{(i)}_{m(0,1),n,2} + j\underline{d}^{(i)}_{m(0,1),n,2}\right), \\
R^2_{0,n} &= (-1)^{n+1} G_1\left(b_0(n) - j\overline{d}^{(i)}_{m(0,1),n,2} + j\underline{d}^{(i)}_{m(0,1),n,2}\right) \\
&\quad + (-1)^n G_2\left(a_0(n) - j\overline{d}^{(i)}_{m(0,1),n,1} + j\underline{d}^{(i)}_{m(0,1),n,1}\right),
\end{aligned}
$$
$$(9.49)$$

where $R^1_{0,n}$ and $R^2_{0,n}$ represent the received signals on the upper and lower subcarriers of the $(0, n)$th real SFBC code, respectively.

After the space-frequency combiner, we have

$$
\begin{aligned}
\tilde{R}^1_{0,n} &= (G_1)^\star R^1_{0,n} + (-1)^n G_2\left(R^2_{0,n}\right)^\star \\
&= \left(|G_1|^2 + |G_2|^2\right) a_0(n) + \left(|G_1|^2 + |G_2|^2\right) j\overline{d}^{(i)}_{m(0,1),n,1} \\
&\quad + \left(|G_1|^2 - |G_2|^2\right) j\underline{d}^{(i)}_{m(0,1),n,1} \\
&\quad + 2(G_1)^\star G_2 j\underline{d}^{(i)}_{m(0,1),n,2}, \\
\tilde{R}^2_{0,n} &= (G_2)^\star R^1_{0,n} + (-1)^{n+1} G_1\left(R^2_{0,n}\right)^\star \\
&= \left(|G_1|^2 + |G_2|^2\right) b_0(n) + \left(|G_1|^2 + |G_2|^2\right) j\overline{d}^{(i)}_{m(0,1),n,2} \\
&\quad - \left(|G_1|^2 - |G_2|^2\right) j\underline{d}^{(i)}_{m(0,1),n,2} \\
&\quad + 2G_1(G_2)^\star j\underline{d}^{(i)}_{m(0,1),n,1}.
\end{aligned}
$$
$$(9.50)$$

The decision variables could then be computed by taking the real part of Eq. (9.50) as

$$
\begin{aligned}
\hat{a}_{0,n} &= \Re\left\{\tilde{R}^1_{0,n}\right\} \\
&= \left(|G_1|^2 + |G_2|^2\right) a_0(n) + 2\underline{d}^{(i)}_{m(0,1),n,2}\Re\left\{j(G_1)^\star G_2\right\},
\end{aligned}
$$

$$\hat{b}_{0,n} = \Re \left\{ \tilde{R}_{0,n}^2 \right\}$$
$$= \left(|G_1|^2 + |G_2|^2 \right) b_0(n) + 2\underline{d}_{m(0,1),n,1}^{(i)} \Re \left\{ jG_1(G_2)^\star \right\}. \tag{9.51}$$

Substituting Eqs. (9.25), (9.41), (9.43) into Eq. (9.51), we have

$$\hat{a}_{0,n} = \left(|G_1|^2 + |G_2|^2 \right) a_0(n)$$
$$+ 2 \sum_{q \in [-r,r]} (-1)^{q+1} a_0(n+q) \left(jc_{1,q}^0 \right) \Re \left\{ (jG_1)^\star G_2 \right\},$$
$$\hat{b}_{0,n} = \left(|G_1|^2 + |G_2|^2 \right) b_0(n)$$
$$+ 2 \sum_{q \in [-r,r]} (-1)^q b_0(n+q) \left(jc_{1,q}^0 \right) \Re \left\{ jG_1(G_2)^\star \right\}. \tag{9.52}$$

It is observed that $a_0(n)$ and $b_0(n)$ only suffer interference from $a_0(n+q)$ and $b_0(n+q)$, respectively (i.e., there is no interference between the two innermost subcarriers). Based on this observation, a simple equalization scheme is presented to independently equalize the innermost subcarriers.

The L real-valued symbols transmitted on the upper innermost subcarrier are denoted as $\mathbf{A} = [a_0(0), a_0(1), \dots, a_0(L-1)]^T$. According to Eq. (9.52), we have

$$\hat{\mathbf{A}} = \mathbf{DA} + \boldsymbol{\eta}, \tag{9.53}$$

where $\hat{\mathbf{A}} = \left[\hat{a}_0(1), \hat{a}_0(2), \dots, \hat{a}_0(L) \right]^T$, $\boldsymbol{\eta} \in \mathbb{C}^{L \times 1}$ is a zero-mean complex AWGN vector, and $\mathbf{D}$ is given as

$$\mathbf{D} =$$

$$
\begin{bmatrix}
|G_1|^2+|G_2|^2+e_0 & e_1 & \cdots & e_r & & & & \\
& \ddots & & & & & & \\
e_{-r} & \cdots & e_{-1} & |G_1|^2+|G_2|^2+e_0 & e_1 & & \cdots & e_r \\
& e_{-r} & \cdots & e_{-1} & |G_1|^2+|G_2|^2+e_0 & e_1 & \cdots & e_r \\
& & e_{-r} & \cdots & e_{-1} & |G_1|^2+|G_2|^2+e_0 & e_1 & \cdots & e_r \\
& & & & & & \ddots & \\
& & & & & e_{-r} \cdots e_{-1} & |G_1|^2+|G_2|^2+e_0
\end{bmatrix}_{L \times L},
$$

$$\tag{9.54}$$

where $e_q = 2 \times (-1)^{q+1} \left(jc_{1,q}^0 \right) \Re \left\{ j(G_1)^\star G_2 \right\}$, $q \in [-r,r]$. The ZF equalizer could then be employed to recover $\mathbf{A}$ as

$$\tilde{\mathbf{A}} = \left(\mathbf{D}^H \mathbf{D} \right)^{-1} \mathbf{D}^H \hat{\mathbf{A}}. \tag{9.55}$$

L is the number of real-valued symbols that are equalized at the same time. Considering the matrix inversion in the equalizer, the equalization can be done in several times and each time with a suitable L in practice. Note that the ZF equalizer is only given as an example of equalization for the upper innermost subcarrier. Other equalization methods (e.g., LMMSE) could also be employed to recover **A** for better performance.

Similarly, the lower innermost subcarrier can be handled in the same manner as the upper innermost subcarrier.

9.3.4 Simulation Results

For simplicity, we assume that the receiver has knowledge of the channel in the simulations. We employ the PHYDYAS filter [15] as the prototype filter with duration of $4T$, where T is symbol spacing. In the receiver, a simple one-tap ZF equalizer is adopted for channel equalization. The length of CP is 1/8 of the OFDM symbol duration. The parameters of the simulations are given in Table 9.1.

Flat Fading Channel

Fig. 9.7 shows the BER performance of the scheme without the guard subcarrier over the flat Rayleigh fading channel. The performances of a conventional SFBC-OFDM system and a single antenna OQAM/FBMC system are presented for comparison. The SFBC-OFDM system performs SFBC coding over adjacent subcarriers, which is different from the block coding in the block-wise SFBC-based OQAM/FBMC system. It is observed that, the performance of the scheme is significantly improved by the ZF equalization, and is very close to that of the conventional SFBC-OFDM scheme. When the block size $K_t = 8$, the scheme with the ZF equalization achieves almost the same performance with the SFBC-OFDM scheme. When K_t decreases, the performance of the scheme with the ZF equalization is slightly degraded. The reason is that the ZF equalization

Table 9.1 Parameters for simulation

Configuration of the antennas	2×1
Number of subcarriers	256
Modulation	16-QAM
Prototype filter	PHYDYAS
Subcarrier spacing	15 kHz
Sampling frequency	3.84 MHz
Frame length	60 Real symbols

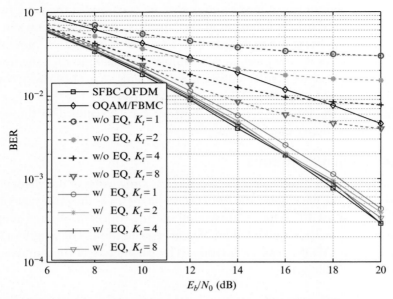

Fig. 9.7 BER performance without guard subcarrier over Rayleigh fading channel.

tends to amplify the noise on the two innermost subcarriers and deteriorates the performance. Hence, the performance is the worst when $K_t = 2$.

Frequency-Selective Fading Channel

In this section, simulations are conducted under three different frequency-selective fading channel models given in *3GPP TS 36.104* [19], namely the Extended Pedestrian A model (EPA), the Extended Vehicular A model (EVA), and the Extended Typical Urban model (ETU).

Due to the variation of channel responses over different subcarriers, the assumption that the channel frequency response does not vary over the subcarriers of the code is no longer suitable. This could result in a performance degradation of the SFBC scheme, especially the block-wise SFBC scheme. Assuming unequal channel responses on the two subcarriers, Eq. (9.28) becomes

$$
\begin{bmatrix} R_1 \\ R_2^\star \end{bmatrix} = \begin{bmatrix} G_{11} & G_{21} \\ G_{22}^\star & -G_{12}^\star \end{bmatrix} \begin{bmatrix} S_1 \\ S_2 \end{bmatrix}, \tag{9.56}
$$

where G_{nm} is the channel frequency response for mth subcarrier at the antenna n. We employ a simple method to eliminate the intersymbol interference (ISI) at the space frequency combiner as

$$\begin{bmatrix} \tilde{R}_1 \\ \tilde{R}_2 \end{bmatrix} = \begin{bmatrix} G_{12}^{\star} & G_{21} \\ G_{22}^{\star} & -G_{11} \end{bmatrix} \begin{bmatrix} R_1 \\ R_2^{\star} \end{bmatrix}$$

$$= \begin{bmatrix} G_{12}^{\star} & G_{21} \\ G_{22}^{\star} & -G_{11} \end{bmatrix} \begin{bmatrix} G_{11} & G_{21} \\ G_{22}^{\star} & -G_{12}^{\star} \end{bmatrix} \begin{bmatrix} S_1 \\ S_2 \end{bmatrix}$$

$$= \begin{bmatrix} G_{11}G_{12}^{\star} + G_{21}G_{22}^{\star} & 0 \\ 0 & G_{11}G_{12}^{\star} + G_{21}G_{22}^{\star} \end{bmatrix} \begin{bmatrix} S_1 \\ S_2 \end{bmatrix}. \quad (9.57)$$

It is clear that Eq. (9.57) does not ensure the full diversity gain especially when the difference between the channel responses of different subcarriers becomes significant. However, it does eliminate the ISI between the two data symbols.

Fig. 9.8 depicts the BER performance of the scheme without the guard subcarrier over the EPA channel. It is observed that the BER performance of the scheme is nearly the same as that in the flat Rayleigh fading channel due to the mild frequency selectivity of the EPA channel.

Fig. 9.9 shows the BER performance of the scheme without the guard subcarrier over the EVA channel. It is observed that the performance of the schemes are degraded due to the frequency selectivity. At BER of

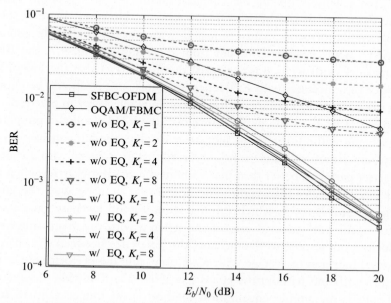

Fig. 9.8 BER performance without guard subcarrier over the EPA channel.

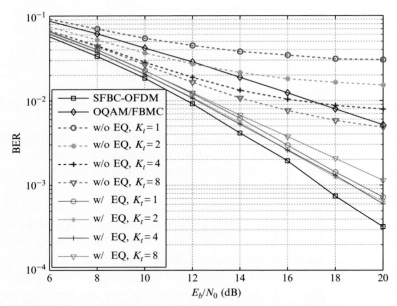

Fig. 9.9 BER performance without guard subcarrier over the EVA channel.

10^{-3}, the scheme with the ZF equalization and $K_t = 4$ is about 1.2 dB worse than that of the SFBC-OFDM scheme. However, there is still a significant performance gain compared with that of the single antenna OQAM/FBMC system. Different from the result in Fig. 9.7, the BER performance with $K_t = 8$ is the worst due to the fact that the loss of the diversity gain increases with K_t. Hence, the block size K_t should be chosen with the consideration of the channel frequency selectivity in practice.

In Fig. 9.10, the BER performance over the ETU channel is presented. The scheme without the guard subcarrier employs the ZF equalization on the two innermost subcarriers. It is observed that the scheme with the guard subcarrier performs better than that without the guard subcarrier when the block size is smaller than eight. The best one is the case with the guard subcarrier and $K_t = 1$, which still maintains a good diversity gain. As K_t increases, the performance gain of the scheme with the guard subcarrier gradually vanishes due to the fact that the difference between the channel responses increases in average with K_t. Comparing the performance of the schemes with and without the guard subcarrier, the following observation is obtained: the simple equalization method is no longer valid in such severe frequency selectivity because the assumption on similar frequency

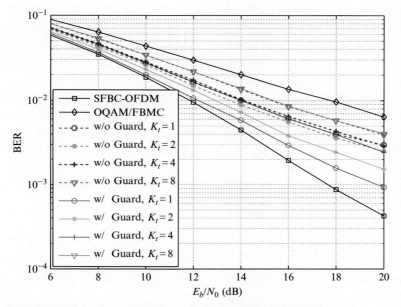

Fig. 9.10 BER performance without guard subcarrier over the ETU channel.

responses over adjacent subcarriers no longer holds. Therefore, satisfactory performance cannot be achieved and this is one of the major factors that restrict the performance of the scheme without the guard subcarrier.

9.4 FINER SVD-BASED BEAMFORMING FOR OQAM/FBMC SYSTEMS

In this section, a spatial diversity technology is introduced. We will introduce a spatial multiplexing technology. We present an architecture with finer beamforming for OQAM/FBMC, here finer beamforming means beamforming with finer granularity in frequency domain. To enable a finer granularity, we adopt the FS-OQAM/FBMC structure, and beamforms at each tone of FS-OQAM/FBMC. The finer beamforming provide a base architecture to support better smoothness between adjacent beamforming matrices.

9.4.1 Frequency Spreading OQAM/FBMC (FS-OQAM/FBMC)

The FS-OQAM/FBMC structure, a special form of the fast convolution implementation of filter banks, uses frequency spreading/despreading to

implement the time domain filtering in the frequency domain for OQAM/
FBMC systems. In FS-OQAM/FBMC, FFT/IFFT is taken at the length
of αK. Let $H(k)$ denote the FFT of a segment of $h(i)$ in the range of $0 \leq i \leq \alpha K - 1$. If properly designed, $H(k)$ has $2P-1$ nonnegligible tones that
center around the zeroth tone, where P is a positive integer. Let $H_{m,n,n_0}(k)$
denote the FFT of a segment of $h_{m,n}(i)$ in the range of $n_0 K/2 \leq i \leq n_0 K/2 + \alpha K - 1$, which is the portion of $h_{m,n}(i)$ that falls inside the n_0th
sliding window. The filtering at the nth time index is then implemented
in the frequency domain by spreading the PAM symbols with $H_{m,n,n}(k)$,
which is the FFT of the nonzero part of pulse $h_{m,n}(i)$. Then, the output of
the frequency spreading are fed to the IFFT transformation to obtain the
time domain samples of the nth time index as

$$x_n(i) = \sum_{k=0}^{\alpha K-1} \left[\sum_{m=0}^{K-1} a_m(n) H_{m,n,n}(k) \right] e^{j\frac{2\pi ki}{\alpha K}}, \quad 0 \leq i \leq \alpha K - 1. \quad (9.58)$$

The transmitted time sequence is then obtained by overlapping and
accumulation as

$$x(i) = \sum_{n\in\mathbb{Z}} x_n \left(i - n\frac{K}{2} \right). \quad (9.59)$$

At the receiver, a sliding window is employed to select αK samples
every $K/2$ samples, which is fed to an FFT module. The transmitted PAM
symbols are recovered through equalization and frequency despreading.

9.4.2 Finer SVD Beamforming

Fig. 9.11 presents the transmitter of the finer singular value decomposition (SVD)-OQAM/FBMC architecture, where $\mathbf{V}_k \in \mathbb{C}^{N_t \times L}$ is

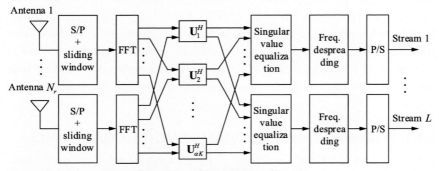

Fig. 9.11 The transmitter of the SVD-OQAM/FBMC architecture.

the beamforming matrix for the kth FS-OQAM/FBMC tone, $\mathbf{V}_k = [\mathbf{v}_k^1 \ \mathbf{v}_k^2 \cdots \mathbf{v}_k^L]$, and $\mathbf{v}_k^l$ denotes the beamforming vector for the lth stream. The beamformed signals on all transmit antennas for the kth tone and nth time index is given by an N_t-by-1 vector

$$\mathbf{b}_n(k) = \mathbf{V}_k \sum_{m=0}^{K-1} \mathbf{a}_{m,n} H_{m,n,n}(k). \tag{9.60}$$

The ith sample (an N_t-by-1 vector) on all transmit antennas for the nth time index is then obtained by IFFT

$$\mathbf{x}_n(i) = \sum_{k=0}^{\alpha K-1} \mathbf{b}_n(k) e^{j\frac{2\pi ki}{\alpha K}}, \quad 0 \le i \le \alpha K - 1. \tag{9.61}$$

After overlapping and accumulation, the transmitted signal on all antennas is represented by the following vector sequence

$$\mathbf{x}(i) = \sum_{n\in\mathbb{Z}} \mathbf{x}_n\left(i - n\frac{K}{2}\right). \tag{9.62}$$

Fig. 9.12 presents the receiver of the SVD-OQAM/FBMC architecture, where $\mathbf{U}_k^H \in \mathbb{C}^{L\times N_r}$ is the beamforming matrix for the kth FS-OQAM/FBMC tone, and $\mathbf{U}_k^H = [(\mathbf{u}_k^1)^H \ (\mathbf{u}_k^2)^H \cdots (\mathbf{u}_k^L)^H]$, where $(\mathbf{u}_k^l)^H$ denotes the receive beamforming vector for the lth stream. Let N_r-by-1 vector $\mathbf{y}(i)$ denote the ith received sample on all receive antennas, and $\mathbf{y}_{n_0}(i)$ ($0 \le i \le \alpha K - 1$) denote the selected αK samples by the n_0th sliding window (i.e., $\mathbf{y}_{n_0}(i) = \mathbf{y}(i + n_0 K/2)$). Assuming that the MIMO channel is nearly flat over the bandwidth of subcarrier, and the transmit beamformers are smooth across the bandwidth, it is obtained after FFT

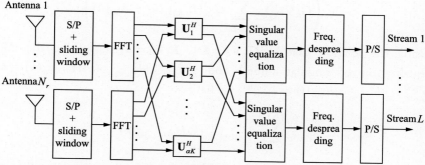

Fig. 9.12 The receiver of the SVD-OQAM/FBMC architecture.

$$\tilde{\mathbf{b}}_{n_0}(k) = \sum_{i=0}^{\alpha K - 1} \mathbf{y}_{n_0}(i) e^{-j\frac{2\pi ki}{\alpha K}}$$

$$\simeq \mathbf{G}_k \mathbf{V}_k \sum_{m=0}^{K-1} \sum_{n \in \mathbb{Z}} \mathbf{a}_{m,n} H_{m,n,n_0}(k) + \mathbf{n}_{n_0}(k), \quad 0 \le k \le \alpha K - 1,$$

$$(9.63)$$

where $\mathbf{n}_{n_0}(k)$ is FFT of the noise with $\mathbf{y}_{n_0}(i)$.

Applying the receive beamformer, we have the signals of all streams on the kth tone

$$\tilde{\mathbf{s}}_{n_0}(k) = \mathbf{U}_k^H \tilde{\mathbf{b}}_{n_0}(k). \qquad (9.64)$$

It should be noted that different tone and stream may have different singular values, therefore an equalizer is required at each tone for each stream. The equalizers for the kth tone is represented by an L-by-L diagonal matrix $\mathbf{E}_k = \text{diag}\{E_k^1, E_k^2, \dots, E_k^L\}$, where E_k^l is the equalizer for Stream l and Tone k. If it is a ZF equalization, $E_k^l = 1/\lambda_k^l$. The equalized signal is $\bar{\mathbf{s}}_{n_0}(k) = \mathbf{E}_k \tilde{\mathbf{s}}_{n_0}(k)$. Finally, despreading is applied to obtain the PAM symbols at the m_0th subcarrier and n_0th time index for all streams

$$\tilde{\mathbf{a}}_{m_0,n_0} = \sum_{k=0}^{\alpha K - 1} H_{m_0,n_0,n_0}^{\star}(k) \bar{\mathbf{s}}_{n_0}(k). \qquad (9.65)$$

Using Eqs. (9.63), (9.64), the estimated PAM symbols at Time n_0 and Subcarrier m_0 is

$$\tilde{\mathbf{a}}_{m_0,n_0} \simeq \sum_{k=0}^{\alpha K - 1} H_{m_0,n_0,n_0}^{\star}(k) \mathbf{E}_k \mathbf{U}_k^H$$

$$\times \left(\mathbf{G}_k \mathbf{V}_k \sum_{m=0}^{K-1} \sum_{n \in \mathbb{Z}} \mathbf{a}_{m,n} H_{m,n,n_0}(k) + \mathbf{n}_{n_0}(k) \right). \qquad (9.66)$$

Using the relation that $\mathbf{U}_k^H \mathbf{G}_k \mathbf{V}_k = \mathbf{D}_k$, $\mathbf{E}_k \mathbf{D}_k = \mathbf{I}_L$ (assuming the ZF equalization of singular values) and $\sum_{k=0}^{\alpha K-1} H_{m_0,n_0,n_0}^{\star}(k) H_{m,n,n_0}(k) = \sum_{i=-\infty}^{\infty} h_{m_0,n_0}^{\star}(i) h_{m,n}(i) = \zeta_{m,n}^{m_0,n_0}$, we have

$$\tilde{\mathbf{a}}_{m_0,n_0} \simeq \mathbf{a}_{m_0,n_0} \sum_{k=0}^{\alpha K - 1} H_{m_0,n_0,n_0}^{\star}(k) H_{m_0,n_0,n_0}(k)$$

$$+ \sum_{(m,n) \neq (m_0,n_0)} \mathbf{a}_{m,n} \sum_{k=0}^{\alpha K - 1} H_{m_0,n_0,n_0}^{\star}(k) H_{m,n,n_0}(k)$$

$$+ \sum_{k=0}^{\alpha K-1} H^{\star}_{m_0,n_0,n_0}(k)\mathbf{E}_k\mathbf{U}_k^H\mathbf{n}_{n_0}(k) \tag{9.67}$$

$$\simeq \mathbf{a}_{m_0,n_0} + \sum_{(m,n)\neq(m_0,n_0)} \mathbf{a}_{m,n} \sum_{k=0}^{\alpha K-1} \zeta_{m,n}^{m_0,n_0}$$

$$+ \sum_{k=0}^{\alpha K-1} H^{z}_{m_0,n_0,n_0} \star (k)\mathbf{E}_k\mathbf{U}_k^H\mathbf{n}_{n_0}(k).$$

Apparently, the first term of Eq. (9.67) is the transmitted PAM symbols, and the second term is the intercarrier interference (ICI)/ISI that is pure imaginary under the nearly flat fading assumption and could be removed by taking the real part.

9.4.3 Smoothing the Beamformers

The finer beamforming architecture itself does not guarantee a low leaked ICI/ISI, it is valid only when the beamformers are smooth across the transmit bandwidth. Therefore, the SVD decomposition to be used in this architecture should be able to provide smooth transition from $\mathbf{V}_{k-1}$ to $\mathbf{V}_k$ for all ks. To satisfy this requirement, we adopt the orthogonal iteration method [20] for SVD decomposition, which was used in [21] to provide smooth beamforming for an OFDM system. As we know, $\mathbf{V}_k$ is the right singular vectors of $\mathbf{G}_k$, as well as the eigenvectors of $\mathbf{A}_k = \mathbf{G}_k^H\mathbf{G}_k$, which can be found by performing the iteration from an initial matrix $\mathbf{Q}^{(0)} \in \mathbb{C}^{N_t \times L}$ with orthonormal columns

$$\mathbf{B}^{(i)} = \mathbf{A}_k\mathbf{Q}^{(i-1)}, \quad i = 1, 2, \ldots$$
$$\text{QR decomposition: } \mathbf{B}^{(i)} = \mathbf{Q}^{(i)}\mathbf{R}^{(i)}, \tag{9.68}$$

where i denotes the iteration times and N_{iter} is the total iteration number. According to [21], $\mathbf{R}^{(i)}$ converges to a diagonal matrix containing the eigenvalues of $\mathbf{A}_k$, and $\mathbf{Q}^{(i)}$ converges to an orthonormal basis for the dominant subspace of dimension L.

Given $\mathbf{V}_{k-1}$, for a beamforming that is smooth from Tone $k-1$ to k, $\mathbf{V}_{k-1}$ could serve as the initial $\mathbf{Q}^{(0)}$ and $\mathbf{V}_k = \mathbf{Q}^{(N_{\text{iter}})}$. It will be shown next that very few iterations are needed to obtain a satisfactory $\mathbf{V}_k$, which at the same time ensures smoothness between $\mathbf{V}_{k-1}$ and $\mathbf{V}_k$. The complete algorithm is presented in Algorithm 9.1.

Algorithm 9.1 Orthogonal Iteration for Smooth Beamforming

Initialize $V_0 = \mathrm{SVD}(G_0)$ // SVD($\cdot$) stands for an arbitrary SVD algorithm;

for $k = 1 : \alpha K - 1$ do

    $A_k = G_k^H G_k$;

    $V_k = V_{k-1}$;

    for $i = 1 : N_{\mathrm{iter}}$ do

        $B_k = A_k V_k$;

        Update V_k using the following QR decomposition: $B_k = V_k R_k$;

    end for

    $D_k = \mathrm{SQRT}(R_k)$ // SQRT($\cdot$) stands for the square root of an diagonal matrix;

end for

9.4.4 Simulation Results

We evaluate the performance of the finer and smooth SVD-based beam-forming for OQAM/FBMC, through computer simulations. We compare the OQAM/FBMC system with an OFDM system, under a setup similar to the IEEE 802.11n wireless LAN standards. The presented results reveal excellent performance of the presented method, which can compete with OFDM and give similar BER results with 64-QAM constellation, under channel model D in the IEEE 802.11n standards.

The following parameters are used for both SVD-OQAM/FBMC and SVD-OFDM systems. The MIMO system is configured as $N_t = N_r = L = 2$. There are $K = 64$ subcarriers, and the subcarrier spacing is 312.5 kHz. Among the 64 subcarriers, there are 48 active subcarriers modulated by QPSK or 64-QAM constellations. For channel coding, we use convolutional codes of rate 2/3 and constraint length 7. A random interleaver is applied after the coding. Each data frame consists of 7 OQAM/FBMC symbol (each consists of 48 OQAM/QAM symbols). The OQAM/FBMC system employs the PHYDYAS filter [17], and the factor α is set to 4. With the FFT size of αK, the filter has seven nonnegligible tones (i.e., $P = 4$). For the singular value equalization, a ZF equalizer is employed at the receiver of SVD-OQAM/FBMC. Then, for the smoothing method, $N_{\mathrm{iter}} = 3$. The SNR in the simulation is defined as: $\mathrm{SNR} \triangleq N_t \sigma_a^2 \sigma_h^2 / \sigma_n^2$, where σ_a^2, σ_h^2, and σ_n^2 are expected signal power of each transmit antenna, expected channel power gain between a pair of transmit and receive antennas, and expected AWGN noise power of each receive antenna, respectively, on each active subcarrier.

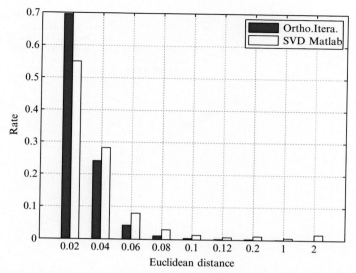

Fig. 9.13 The Euclidean distance between transmit beamformers of adjacent tones.

Let us first check if smoothness across tones is improved with the smoothing method. Smoothness between two beamformers of adjacent tones is evaluated by their differences. Fig. 9.13 presents the histogram of the Euclidean distance between adjacent beamformers by the presented method. Result by the SVD function in the Matlab software is also presented for comparison. The distance of the scheme falls in the range of 0–0.1, while that of SVD Matlab could go as high as 1.0 with nonnegligible percentage. It is thus concluded that the orthogonal iteration method provides a significant smoothness improvement compared with the SVD of no smoothness consideration.

Figs. 9.14–9.16 present the BER performance of the SVD-OQAM/FBMC system with the finer beamforming architecture and smoothing. For comparison, we also present BER results of the following four systems: (1) SVD-OFDM; (2) SVD-OQAM/FBMC with subcarrier-level beamforming but without smoothing (SVD Matlab); (3) SVD-OQAM/FBMC with subcarrier-level beamforming and smoothing (orthogonal iteration); and (4) SVD-OQAM/FBMC with the finer beamforming but without smoothing (SVD Matlab).

The results clearly show that the SVD-OQAM/FBMC system with the finer beamforming and smoothing greatly outperforms the other SVD-OQAM/FBMC systems. It can perform very closely with

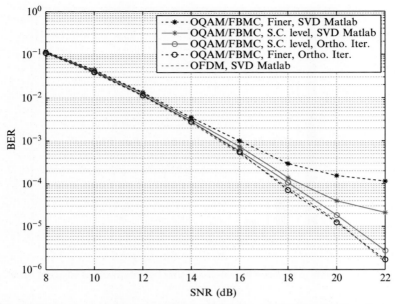

Fig. 9.14 BER performance of the SVD-OQAM/FBMC systems with QPSK and rate 2/3 coding, subcarrier level or finer beamforming, orthogonal iteration or Matlab SVD.

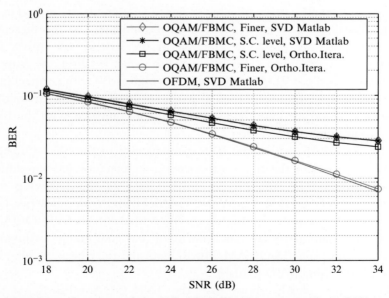

Fig. 9.15 BER performance of the SVD-OQAM/FBMC systems with 64-QAM and no coding, subcarrier level or finer beamforming, orthogonal iteration or Matlab SVD.

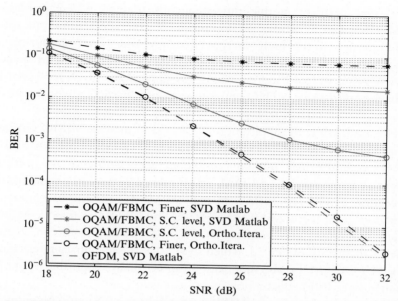

Fig. 9.16 BER performance of the SVD-OQAM/FBMC systems with 64-QAM and rate 2/3 coding, subcarrier level or finer beamforming, orthogonal iteration or Matlab SVD.

SVD-OFDM, in terms of BER, under the IEEE 802.11n channel model D. Be noted that the BER is plotted with respect to SNR in these figures. If we use E_b/N_0 instead of SNR for x-axis, about 1 dB gain of the SVD-OQAM/FBMC will be observed over SVD-OFDM, because that the SVD-OQAM/FBMC does not pay for the energy overhead to transmit the 25% cyclic prefix as the wireless LAN OFDM does.

It is also evident from these results that smoothing is very crucial for the BER performance: both the finer and subcarrier-level SVD-OQAM/FBMC systems with smoothing outperform those without smoothing. The results also show that iterations as few as three is adequate for orthogonal iterations–based SVD. Thus, the SVD based on orthogonal iterations not only results in smooth beamformers but also low complexity due to its incremental approach.

9.5 OQAM/FBMC FOR MASSIVE MIMO

Recently, the massive MIMO system has attracted a lot of attention, where a base station (BS) is equipped with hundreds of antennas and simultaneously serves tens of users in one time-frequency resource [22–25]. Massive MIMO is taken as an enabler for the next generation communications in future due to its high energy efficiency and system performance. Considering the advantages of OQAM/FBMC over OFDM in spectral efficiency and out-of-band emissions, the application of OQAM/FBMC to the area of massive MIMO communications is particularly promising to achieve a higher energy and spectral efficiency.

However, while the application of OFDM to massive MIMO channels is a straightforward task, direct application of OQAM/FBMC to massive MIMO channels may not be possible in general due to the intrinsic imaginary interference of OQAM/FBMC. In [26], the authors discuss the application of OQAM/FBMC to massive MIMO system and present a qualitative comparison of OQAM/FBMC and OFDM in massive MIMO system. *Self-equalization* property of OQAM/FBMC systems in massive MIMO channels is presented, which shows that the imaginary interference caused by OQAM/FBMC modulation can be totally eliminated by simple matched filter (MF) equalization in the receiver.

We consider a multiuser MIMO setup where a BS with N_r antennas simultaneously serves U users in a time division duplexing manner. Each user is equipped with a single antenna and $N_r \gg U$. OQAM/FBMC modulation is used for data transmissions. Ignoring the time and subcarrier indexes of the OQAM/FBMC modulation, the BS received signal vector can be expressed as

$$\mathbf{Y} = \sum_{u=0}^{U-1} [s(u) + jq(u)] \mathbf{g}_u + \mathbf{n}, \tag{9.69}$$

where $s(u)$ is a transmit symbol from the uth user, $q(u)$ is the corresponding ISI and ICI interference due to the OQAM/FBMC modulation, $\mathbf{g}_u = [g_u(0) \cdots g_u(N_r - 1)]^T$ represents the channel vector between the uth user and different antennas at the BS, and $\mathbf{n} \in \mathbb{C}^{N_r \times 1}$ denotes the AWGN. At the receiver, the transmitted symbols are then recovered through the MF detector [26],

$$\hat{s}(u) = s(u) + \sum_{l=0, l \neq u}^{U-1} \frac{\tilde{\mathbf{g}}_u^T}{\|\tilde{\mathbf{g}}_u\|^2} \left[s(l) \tilde{\mathbf{g}}_l + q(l) \breve{\mathbf{g}}_l \right] + \frac{\tilde{\mathbf{g}}_u^T}{\|\tilde{\mathbf{g}}_u\|^2} \tilde{\mathbf{n}}, \tag{9.70}$$

where $\hat{s}(u)$ is the estimated data of user u, $\tilde{\mathbf{g}}_u = \begin{bmatrix} \mathbf{g}_{u,R}^T & \mathbf{g}_{u,I}^T \end{bmatrix}^T$, $\breve{\mathbf{g}}_u = [-\mathbf{g}_{u,I}^T \quad \mathbf{g}_{u,R}^T]^T$, and $\tilde{\mathbf{n}} = [\mathbf{n}_R^T \quad \mathbf{n}_I^T]^T$ with the subscripts R and I denote the real and imaginary parts, respectively. It is evident that the multiuser interference introduced by the channel and the imaginary interference introduced by the OQAM/FBMC modulation tend to vanish when the number of receiving antennas is sufficiently high. Such a property has been called *self-equalization*, which allows one to relax on the requirement of having approximately flat gain for the subcarriers. This, in turn, allows one to reduce the number of subcarriers in an assigned bandwidth. Reducing the number of subcarriers leads to a lower complexity, sensitivity to carrier frequency offset and system latency, etc., which all position OQAM/FBMC as a strong candidate in the application of massive MIMO.

However, the current extension of OQAM/FBMC to massive MIMO system is derived assuming a time flat channel, which is not eligible in the real environment with relatively large channel delay spread. In the broadband systems, it is conventional that multitap equalizers have to be employed in OQAM/FBMC for channel equalization [27], which has a high computational complexity and may not work well in massive MIMO systems.

To enable the simple single-tap equalizer to work well, a different OQAM/FBMC system with cyclic prefix, denoted as CP-OQAM/FBMC, has been proposed recently [28]. Similar to OFDM systems, a CP-OQAM/FBMC system converts a linear convolutional channel to a circular convolutional channel and makes it possible to employ the simple single-tap equalizer for the channel equalization in the frequency domain. Therefore, the application of CP-OQAM/FBMC to the massive MIMO systems is directly at the cost of spectral efficiency, which is similar to conventional OFDM. The exploration of OQAM/FBMC techniques with or without the insertion of some redundant CP in the area of massive MIMO is a new research direction and more researches shall be developed on it.

9.6 SUMMARY

In this chapter, we presented a block-wise SFBC scheme for OQAM/FBMC systems with a guard subcarrier to exploit the diversity gain in MIMO OQAM/FBMC, which guarantees the orthogonality condition of the Alamouti code. To improve the spectral efficiency, we also considered the case without the guard subcarrier. Since only the two innermost subcarriers do not satisfy the complex orthogonality condition of the Alamouti code, a simple equalization scheme was presented to independently equalize the two innermost subcarriers. Simulation results showed that the presented scheme works well over channels with mild to moderate frequency selectivity. Then, to exploit the beamforming gain in MIMO OQAM/FBMC, a finer SVD-based beamforming scheme for OQAM/FBMC systems was presented, which greatly improves the smoothness of beamformers, therefore effectively suppresses the leaked ICI/ISI from adjacent subcarriers. Simulation results showed that the SVD-FBMC/OQAM system shares very close BER performance with its OFDM counterpart under the frequency selective channels. Also, it is worthwhile to note that the application of OQAM/FBMC to the area of massive MIMO communications is particularly promising to achieve higher energy and spectral efficiency, and more researches shall be developed on it.

REFERENCES

[1] Caus M, Pérez-Neira AI. Transmitter-receiver designs for highly frequency selective channels in MIMO FBMC systems. IEEE Trans Signal Process 2012;60(12):6519–32.
[2] Bellanger M. FBMC physical layer: a primer; 2010. Available from: http://www.ict-phydyas.org/teamspace/internal-folder/FBMC-Primer_06-2010.pdf.
[3] Paulraj AJ, Gore DA, Nabar RU, Bolcskei H. An overview of MIMO communication: a key to gigabit wireless. IEEE Proc 2004;92(2):198–218.
[4] Alamouti SM. A simple transmit diversity technique for wireless communications. IEEE J Sel Areas Commun 1998;46(4):1451–8.
[5] Tse D, Viswanath P. Fundamentals of wireless communication. New York, NY: Cambridge University Press; 2005.
[6] Lele C, Siohan P, Legouable R. The Alamouti scheme with CDMA-OFDM/OQAM. EURASIP J Adv Signal Process 2010;2010(2):1–13.
[7] Ihalainen T, Ikhlef A, Louveaux J, Renfors M. Channel equalization for multi-antenna FBMC/OQAM receivers. IEEE Trans Veh Technol 2011;60(5):2070–85.
[8] Renfors M, Ihalainen T, Stitz TH. A block-Alamouti scheme for filter bank based multicarrier transmission. In: European wireless conference, Lucca; 2010.
[9] Hao L, Lele C, Siohan P. A pseudo Alamouti transceiver design for OFDM/OQAM modulation with cyclic prefix. In: IEEE 10th workshop on signal processing advances in wireless communications, Perugia; 2009.
[10] Zakaria R, Ruyet DL. On interference cancellation in Alamouti coding scheme for filter bank based multicarrier systems. In: 10th international wireless communication systems; 2013.
[11] Seshadri N, Winters JH. Two schemes for improving the performance of frequency-division duplex (FDD) transmission systems using transmitter antenna diversity. Int J Wirel Inf Netw 1994;1:49–60.
[12] Tarokh V, Seshadri N, Calderbank AR. Space-time codes for high data rate wireless communication: performance criteria and code construction. IEEE Trans Inf Theory 1998;44(2):C744–65.
[13] Jafarkhani H. A quasi-orthogonal space-time block code. IEEE Trans Commun 2001;49(1):1–4.
[14] Tarokh V, Jafarkhani H, Calderbank AR. Space-time block codes from orthogonal designs. IEEE Trans Inf Theory 1999;45(5):1456–67.
[15] Bellanger MG. Specification and design of a prototype filter for filter bank based multicarrier transmission. In: IEEE international conference on acoustics, speech, and signal processing; 2001.

[16] Mirabbasi S, Martin K. Overlapped complex-modulated transmultiplexer filters with simplified design and superior stopbands. IEEE Trans Circuits Syst II Analog Digit Signal Process 2003;50(8):456–69.

[17] Viholainen A, Bellanger M, Huchard M. Prototype filter and structure optimization, 2009. Available from: http://www.ict-phydyas.org/delivrables/PHYDYAS-D5-1.pdf/view.

[18] Medjahdi Y, Terre M, Ruyet DL, Roviras D. On spectral efficiency of asynchronous OFDM/FBMC based cellular networks. In: IEEE international symposium on personal indoor and mobile radio communications, Toronto; 2011.

[19] 3GPP TS 36.104 V10.2.0. Base station (BS) radio transmission and reception; 2011.

[20] Golub G, Van Loan C. Matrix computations. Baltimore, MD: Johns Hopkins University Press; 1996.

[21] Sandell M, Ponnampalam V. Smooth beamforming for OFDM. IEEE Trans Wirel Commun 2009;8(3):1133–8.

[22] Larsson EG, Edfors O, Tufvesson F, Marzetta T. Massive MIMO for next generation wireless systems. IEEE Commun Mag 2014;52(2):186–95.

[23] Hgo HQ, Larsson EG, Marzetta TL. Energy and spectral efficiency of very large multiuser MIMO systems. IEEE Trans Commun 2013;61(4):1436–49.

[24] Rusek F, Persson D, Buon KL, Larsson EG. Scaling up MIMO: opportunities and challenges with very large arrays. IEEE Signal Process Mag 2013;30(1):40–60.

[25] Marzetta TL. Noncooperative cellular wireless with unlimited numbers of base station antennas. IEEE Trans Wirel Commun 2010;9(11):3590–600.

[26] Farhang A. Filter bank multicarrier for massive MIMO. VTC Fall; 2014.

[27] Waldhauser DS, Baltar LG, Nossek JA. Adaptive equalization for filter bank based multicarrier systems. In: IEEE international symposium on circuits and systems; 2008.

[28] Gao X, Wang W, Xia XG, Au EKS, You X. Cyclic prefixed OQAM-OFDM and its application to single-carrier FDMA. IEEE Trans Commun 2011;59(5):1467–80.

CHAPTER 10

Applications

With the continuous arising of new mobile Internet applications, the future wireless communication systems are surely required to be massive and tactile interconnected compared to conventional wireless communication systems. However, current physical layer techniques for wireless communications, such as conventional orthogonal frequency division multiplexing (OFDM)-based multicarrier communications, are very difficult to support typical applications in massive-connection and tactile Internet. The main reason is that conventional OFDM systems employ time-domain rectangular window with very poor frequency localization, resulting in that communication nodes need to consume much time for signaling overhead and transmission waiting. In addition, it is very hard to keep strict orthogonality by adjusting system parameters flexibly, such as subcarrier frequency and spacing, which means that conventional OFDM systems are very difficult to be employed in massive-connected environment in future wireless communications with diversified services.

Recently, offset quadrature amplitude modulation-based filter bank multicarrier (OQAM/FBMC) has been considered as a promising alternative to the conventional OFDM technique, especially for coordinated multipoint (CoMP), fragmented spectrum, machine-type communication (MTC), and tactile interconnected communication scenarios. OQAM/FBMC has a wide application in future 5G communications [1, 2]. In this chapter, we will present several potential applications of 5G scenarios employing OQAM/FBMC techniques.

The rest of this chapter is organized as follows. In Section 10.1, we describe the CoMP transmission OQAM/FBMC systems. Fragmented spectrum is presented in Section 10.2. Later, MTC and tactile interconnected communication scenarios are discussed in Sections 10.3 and 10.4, respectively. Finally, summary is made in Section 10.5.

OQAM/FBMC for Future Wireless Communications
http://dx.doi.org/10.1016/B978-0-12-813557-0.00010-3
© 2018 Elsevier Ltd.
All rights reserved. **247**

10.1 COORDINATED MULTIPOINT TRANSMISSION

CoMP has been considered to be one typical 5G transmission scenario. Theoretical work has shown that CoMP offers the potential to increase throughput significantly [3–8], in particular at the cell edge, which leads to enhanced fairness overall. However, it is noted that the multiuser CoMP joint reception case where we face timing and carrier frequency offsets is an important scenario for 5G [2]. Timing advance control mechanisms in LTE can align receive signals of a single cell, but, due to propagation delay differences, CoMP joint reception across multiple cells has inherent timing offsets degrading the potential performance gains of current LTE-A. Moreover, relaxed oscillator requirements (like those in WLAN instead of LTE) are beneficial in terms of cost and performance for mass market devices. Thus, traditional OFDM CoMP schemes are not applicable for 5G.

Due to the use of pulse-shaping filters, the OQAM/FBMC technique does not require strict orthogonality and synchronization, which saves lots of communication resources and time overhead, especially for the scenario involving multiple nodes. Therefore, OQAM/FBMC technique is suitable to support the CoMP transmission. Currently, existing schemes about OQAM/FBMC-CoMP are rare. In this section, we introduce two important and specific scenarios of uplink (UL) and downlink (DL) CoMP for 5G, where OQAM/FBMC is promising and OFDM is limited.

The UL CoMP with joint reception is illustrated in Fig. 10.1, which comprises two cells and two users with timing and frequency offsets.

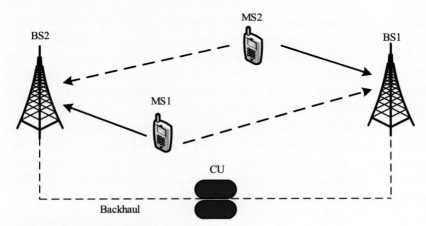

Fig. 10.1 Uplink CoMP reference scenario.

Sources of timing offsets can be the following. UL timing advance control mechanisms can align receive signals of a single cell. However, due to propagation delay differences, UL CoMP joint reception across multiple cells has inherent timing offsets. At the receiver, there also exist the cumulative effects of propagation delay, synchronicity, and delay spread. Regarding frequency offset remarks, the total frequency offsets are caused by both transmitter and receiver. It has to be checked whether relaxed oscillator requirements (like those in WLAN instead of LTE) are significantly beneficial for the price of cheap devices (e.g., as a mass of small "Internet-of-things" sensor-like devices have to be supported by 5G). This would push multicarrier waveforms with lower side lobe levels, namely OQAM/FBMC technique, as higher frequency offsets could be tolerated due to the reduced ICI.

As illustrated in Fig. 10.2, the scenario for OQAM/FBMC deals with DL CoMP with joint reception, where two cells and two users are included and timing and frequency offsets are considered. On the one hand, received signals from multiple cells may not be aligned in time at the receiver due to the difference of propagation delay and the possible time desynchronization between cooperating BSs. The timing and frequency offsets would cause pilots rotations at the receiver and make the channel estimation more difficult. On the other hand, clocks at the BSs and at the UE sides may not be perfectly synchronized in frequency, causing CFO at the receiver. Therefore, the impacts of the time offset and CFO should be studied, as

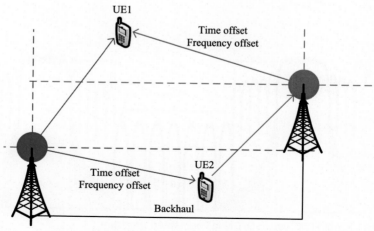

Fig. 10.2 Downlink CoMP reference scenario.

well as the level of tolerance to asynchronism in the DL CoMP scenario using the OQAM/FBMC waveform.

10.2 FRAGMENTED SPECTRUM

Fragmented spectrum may be viewed as the consequence of relaxed extension of channel aggregation for wireless communications. A user may access more than one frequency band being contiguous or not in frequency at any given time. In the asynchronous scenario of 5G, multiple users may not necessarily be synchronized in time, which leads to the generation of a heavily fragmented spectrum both in time and frequency.

As is well known, spectrum is becoming a very rare communication resource. How to exploit the spectrum more efficient is an interesting academic issue and can benefit hugely. One of the solutions is the physical layer technique with low spectrum sidelobe. As shown in Fig. 10.3, both OFDM and OQAM/FBMC may theoretically be suited to multicarrier-based spectrum pooling. However, high adjacent channels rejection cannot be met without a very complex and programmable band-pass transmit filter in the OFDM case, whereas OQAM/FBMC would simply require "switching on and off" the appropriate carriers at the transmitter. The main shortcoming of the OFDM waveform identified here originates from the large sidelobes because of the rectangular shaping of the temporal signal whereas the OQAM/FBMC built-in filtering feature (100× better localization) adapts to spectrum availability even in the fragmented case.

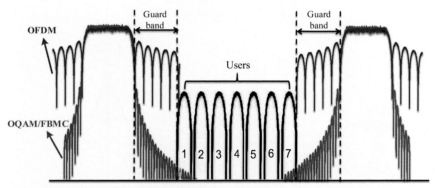

Fig. 10.3 OQAM/FBMC versus OFDM in fragmented spectrum.

Scenarios of fragmentation should be addressed, where a single user is allocated a pool of frequencies with a relaxed synchronization in time. This scenario should have strong implications for the receiver architecture of the BSs as the different users may not be synchronized. A specific synchronization per user should be explored while containing the complexity of the overall receiver in order to avoid unrealistic implementation proposals. The study could take as a starting point of the channelization proposed by current LTE standards, while adapting it to the MTC scenarios. The impact of channel bonding schemes (carrier aggregation vs. spectrum pooling) must be studied at different levels including peak-to-average power ratio requirements, emitted spectrum adjacent channel leakage and requirements on finite precision dynamic and digital-to-analog conversion. Therefore, the added value of the hardware implementation is to provide realistic limits of what can be expected from a fragmented scheme.

10.3 MACHINE-TYPE COMMUNICATION

MTC has been the main application form of Internet of things at present as well as in the future, with numerous application scenarios, such as intelligent transportation, intelligent measurement and control, electronic medical, and so on. According to the European Telecommunications Standard Institute architecture, the MTC system is composed of three domains (i.e., the MTC component working in the device domain, the MTC area network and gateway in the network domain, and the MTC server and communication network in the application domain). Fig. 10.4 demonstrates an example of a feasible MTC network architecture. In this architecture, some single devices equipped with its own subscriber identity module (SIM) card are connected to the core networks through the access network, while other devices may create MTC area networks locally and get connected to the core networks through MTC gateways.

Recent researches mostly focus on how to enhance the low-cost MTC terminal, achieve ultra real-time access, and optimize transmission efficiency. The evolution direction of MTC includes massive terminal access, which means that the number of terminals in a cell is far beyond the normal access level. A large number of devices in a local area transmit data simultaneously in a short time. It requires that access blocking and transmission congestion do not occur in the network. At the same time,

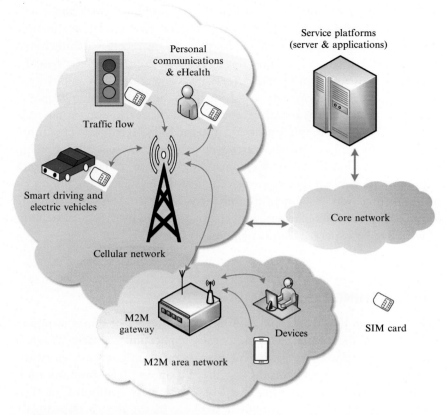

Fig. 10.4 Example of MTC network architecture.

smart grid/intelligent industry, intelligent transportation, remote medical treatment, and other scenarios have a demand of ultra real time.

With a large number of users and a small amount of data, to maintain synchronization and orthogonal of the whole network is not only economic, but also unrealistic in MTC. Moreover, the huge synchronization overhead is also a threat to network stability. The traditional OFDM technique uses a rectangular window in time domain and its localization in frequency domain is poor. In order to ensure the orthogonality between the carriers, the multicarrier signals of different users require strict synchronization. This results in MTC such a large number of communication nodes spending a lot of time in the signaling overhead and transmission waiting. Therefore, it is difficult to meet the requirement of low latency.

OQAM/FBMC uses prototype filters with lower out-of-band leakage on every subcarrier. Hence, time and frequency synchronization requirements are lower than that of OFDM. Therefore, OQAM/FBMC is more suitable for asynchronous transmission.

10.4 TACTILE INTERNET

With the continuous emergence of new mobile Internet applications, compared with traditional wireless communication systems, future wireless communication systems will support two new features: giant link and mobile tactile Internet. On the one hand, CISCO predicts that by 2019, there will be more than 11 billion mobile intelligent terminals getting access to the Internet through mobile communication systems simultaneously, which shows as giant link. On the other hand, control and perception-based electronic medical (delay is less than 10 ms), intelligent manufacturing (delay less than 5 ms), intelligent transportation (delay is less than 1 ms) and other mobile tactile Internet applications demand will gradually occupy the dominant position. The term "tactile Internet" refers to applications based on control communications (through cellular networks) that provide the capability of engaging and steering real and virtual objects at real-time interaction speeds. A short round-trip latency of approximately 1 ms is the key requirement of tactile Internet, which brings a huge challenge for the transmission techniques in physical layer.

However, the physical layer of the existing mobile communication system is mainly based on the traditional OFDM system, which is difficult to meet the requirements of tactile Internet. It is because that the traditional OFDM technique uses a rectangle window in time domain and its localization in frequency domain is poor. Therefore, the multicarrier signals of different users need strict synchronization, which results in a large number of communication nodes spending a lot of time in the signaling overhead and transmission waiting. Compared to the traditional OFDM system, the OQAM/FBMC system adopts more flexible filter bank to shape the transmitted signal and increase its frequency localization. The OQAM/FBMC system is more suitable for asynchronous transmission, and thus avoid the synchronization of waiting and signaling overhead, which can achieve the goal of reducing the latency.

In addition, the OQAM/FBMC system has good frequency localization, low side-lobe spectrum, and good ability of resisting narrow band interference. Hence, OQAM/FBMC could enhance the probability for a successful transmission, and decrease the latency caused by retransmission. OQAM/FBMC shows high reliability and can meet the requirements of the future tactile Internet.

10.5 SUMMARY

In this chapter, we discussed the prospect of OQAM/FBMC in four applications (i.e., CoMP, fragmented spectrum, MTC, and tactile Internet). It is noted that all of them have a strict requirement on synchronism and orthogonality, which cannot be achieved easily. Thus, our idea is to abandon synchronism and orthogonality altogether, thereby admitting some crosstalk or interference, and to control these impairments by OQAM/FBMC, a suitable transceiver structure and transmission technique. In conclusion, OQAM/FBMC is a promising technique for 5G and has a significant research value.

REFERENCES

[1] Farhang-Boroujeny B. OFDM versus filter bank multicarrier. IEEE Signal Process Mag 2011;28(3):92–112.
[2] 5GNOW, 5G Waveform Candidate Selection; 2013. Available from: http://www. 5gnow.eu/wp-content/uploads/2015/04/5GNOW_D3.1_v1.1_1.pdf.
[3] Marsch P, Khattak S, Fettweis G. A framework for determining realistic capacity bounds for distributed antenna systems. In: Information theory workshop; 2006.
[4] Karakayali MK, Foschini GJ, Valenzuela RA. Network coordination for spectrally efficient communications in cellular systems. IEEE Wirel Commun 2006;13(4):56–61.
[5] Parkvall S, Dahlman E, et al. LTE-advanced-evolving LTE towards IMT-advanced. VTC Fall; 2008.
[6] Caire G, Shamai S. On the achievable throughput of a multiantenna Gaussian broadcast channel. IEEE Trans Inf Theory 2003;49(7):1691–706.
[7] Irmer R, Droste H, Marsch P. Coordinated multipoint: concepts, performance, and field trial results. IEEE Commun Mag 2011;49(2):102–11.
[8] Muller A, Frank P. Cooperative interference prediction for enhanced link adaptation in the 3GPP LTE uplink. In: Vehicular technology conference; 2010.

INDEX

Note: Page numbers followed by *f* indicate figures, and *t* indicate tables.